Foot Surgery Viewed Through the Prism of Comparative Anatomy

Comparative Anatomy and Posture of Animal and Human Set

coordinated by
Djillali Hadjouis

Volume 1

Foot Surgery Viewed Through the Prism of Comparative Anatomy

From Normal to Useful

Cyrille Cazeau

ISTE

WILEY

First published 2020 in Great Britain and the United States by ISTE Ltd and John Wiley & Sons, Inc.

ISTE Ltd
27-37 St George's Road
London SW19 4EU
UK

www.iste.co.uk

John Wiley & Sons, Inc.
111 River Street
Hoboken, NJ 07030
USA

www.wiley.com

Library of Congress Control Number: 2020940426

British Library Cataloguing-in-Publication Data
A CIP record for this book is available from the British Library
ISBN 978-1-78630-604-3

MIX
Paper from responsible sources
FSC® C013604

Contents

Foreword

The fact that an orthopedic surgeon, a foot specialist, integrates comparative anatomy into his or her surgical procedures to repair an impaired function is not self-evident. For any paleoanthropologist accustomed to analyzing the bone fragments of our ancestors and their locomotor system, comparative anatomy is common practice; it is an essential point of view from which we base our functional interpretations and evolutionary hypotheses. It is therefore with immense pleasure, a certain pride and not without trepidation that I accepted Dr Cyrille Cazeau's invitation to introduce his work, by evoking some convergences between the comparative approach and the object of study. While it is indeed comparative anatomy and locomotor function that brought us together, comparative anatomy – which Cyrille Cazeau has been working with for many years – gives it a central place in his teachings and workshops, questions it and nourishes it with encounters, such as the one with Pau Golano, an exceptional anatomist whom Cyrille Cazeau introduced me to and I got to know better, to the point of integrating it into his practice of percutaneous foot surgery. For the surgeon, comparing is to understand the human specificity, to distinguish what is essential to the function – useful – from what is simply accessory and, if necessary, can be sacrificed. My approach to the fossil hominid foot, which is unique because it is often fragmented and atypical in shape, owes much in its construction to my encounter with Cyrille Cazeau. This normal and useful element that he questions in his practice has indeed become the central notion in my practice as a paleoanthropologist when it comes to functionally interpreting locomotor anatomy that is often very partial, whereas the only well-known and usable bipedal reference system was the human functional anatomy model, which is so particular. It is in response to this questioning that we sought to study other bipedal reference frames, more particularly non-human primates. So for us, comparing is also understanding non-human specificities. In our approaches to a paleontological object, we now see that bipedalism is not the proper domain of man: it is carried out in a varied manner in our closest cousins, catarrhine primates, whose anatomy seems to be "made" for other mechanically different locomotor modes. This is also how we understand that other efficient (and not transitional) bipeds could have existed in our distant ancestors, relying on the numerous paleoanthropological discoveries of the last two decades and the unexpected anatomical variation that they have revealed in hominids over the last 7 million years. At the time of publication, the discovery of a new primate species, *Danuvius guggenmosi*, in the Hammerschmiede deposit in present-day Bavaria, provides evidence of a potentially much older form of bipedalism. Its age – 11.6 million years – before the supposed differentiation date of the first hominids (one of the commonly accepted features of which was the practice of a habitual bipedalism), drives us to continue this comparative work to better understand human and non-human specificities. Beyond the paleontological object, the functional interpretation that is proposed and its evolutionary implications, it is essential that our comparative and paleoanthropological work be useful. The fact that this usefulness manifests itself in the daily life of each person is most often a very distant perspective, rarely palpable. With *Foot Surgery Viewed Through the Prism of Comparative Anatomy*, Cyrille Cazeau shows us that this perspective is not so distant and, in so doing, invites those working in fundamental and applied research, anatomists, zoologists, physiologists, biomechanists, orthopedists, surgeons, paleoanthropologists, etc. to engage in more discussion to achieve something useful. Finally, Cyrille Cazeau brings together multiple scales, from macroscopic anatomical structures to the bone's microstructure, from millions of years of evolution to the "instantaneous" surgical procedure, towards an application that aims to improve an impaired function. This book, which questions the notion of an anatomical standard, to give full scope to the useful and functional, is therefore not a normal book. It is a book that will be useful to everyone in his or her practice.

Gilles BERILLON
Paleoanthropologist
Director of Research at the CNRS
Laboratoire Histoire naturelle de l'Homme Préhistorique
Musée de l'Homme, Paris
July 2020

Introduction

Orthopedic surgery aims to improve the function of the musculoskeletal system. It aims to repair, rebuild or replace damaged anatomical elements with a prosthesis.

Physical actions in orthopedic surgery always lead to anatomical sacrifices, the benefits of which must outweigh the disadvantages.

To do this, we must differentiate between the anatomical elements that constitute our original anatomical normality and the elements that are truly useful for carrying out a given function, such as permanent terrestrial human bipedalism.

To help differentiate between "normal" and "useful" elements, it is useful to compare similar functional anatomy models, whose differences highlight the specificity of each.

To illustrate this approach, we take the study of the foot as an example. The human model is compared to those with the closest phylogeny: non-human primates and fossil hominids.

Thus, studying anatomical homologies allows us to rank the different anatomical players according to their usefulness for a given function. Some considered as "normal" define the anatomical normality acquired during ontogenesis, while others, essential to achieving a function, are described as "useful".

The surgical consequences are direct in the choices made to correct a forefoot deformity such as hallux valgus. A choice means determination, and determination is negation. We thus implicitly deny everything that does not fall within its definition. Surgery requires that we make a determination that we decide on a procedure, an action by rejecting other possible ones. This is where the quest for the most efficient, the most correct procedure to achieve the best possible result lies, in this interaction between determination and negation.

Classical surgery mainly focuses on restoring *ad integrum* defining original normality. The study of comparative functional anatomy helps to define surgery that will restore the specific useful function, even if it is accompanied by anatomical sacrifices. Therefore, the preliminary work consists of dividing and then ranking the usefulness of the anatomical players for a given species, in this case humans.

The purpose of the surgical act, which always constitutes a physical aggression, a transgression, is to propose in return a result leading, according to Aristotle, the patient as close as possible to recovery. To do this, the injured organ must be put back into its usual working context as soon as possible. We propose a concept – that of percutaneous surgery – composed of linked and inseparable elements. It is advisable to correct the deformations in the three planes of space through minimal surgical approaches, to offer reliable mechanical stability before any biological consolidation allowing immediate complete support and to use an anesthesia technique that avoids the painful postoperative storm which follows traditional techniques.

The benefit, therefore, is faster functional recovery, which is vital for a better outcome.

1

Comparison: The Key to Understanding Human Specificity

1.1. Comparison, a source of clarity

It is difficult to understand the specificity of an object or a function, if we do not compare it with a model that resembles it. Highlighting differences will allow us to grasp their distinctiveness. The smaller the difference, the more difficult it is to detect if it is isolated from a similar environment, but the more it characterizes the specificity of the object being studied.

This easy-to-understand approach when comparing color, from blue to green via emerald, for example, is more complex when it comes to functional anatomy models. Our understanding of specificity cannot be dissociated from the effort of classification and the establishment of phylogeny.

Indeed, by carefully establishing derived and ancestral characteristics, it is possible to establish differences and then classify according to the principle of parsimony. According to Aristotle, let us "search for relations between apparently independent things and search for similarities between things that are dissimilar in the eyes of the common man". Knowing things makes it possible to see them acutely; verbalizing them makes it possible to think about them.

1.2. Different models of the foot and ankle throughout evolution

Locomotion involves a variety of movements adapted to different environments and lifestyles.

Identical morphological organization and processes common to all vertebrates are implemented, but thanks to original anatomical elements, animals can move with highly sophisticated means and respond, each in their own way, to the constraints of their environment.

Among the many elements involved in the musculoskeletal system's function, the foot and ankle (crurotarsal or talocrural joint) were chosen because they undergo the most spectacular and elaborate transformations during adaptation to different types of locomotion: running, trotting, tree climbing, etc. Mammals have anatomical adaptations that reflect their specific mode of locomotion.

1.2.1. *Evolution of the chiridial limb*

The chiridial limb represents the general limb structure of tetrapods. There are significant differences between mammalian classes, but the general pattern is identical, as shown in Figure 1.1.

The foot represents the autopod (Figure 1.1); succeeding the leg sector which constitutes the zeugopod. The autopod consists of three parts: the tarsus (basipod), the metatarsus (metapod) and the phalanxes (acropod).

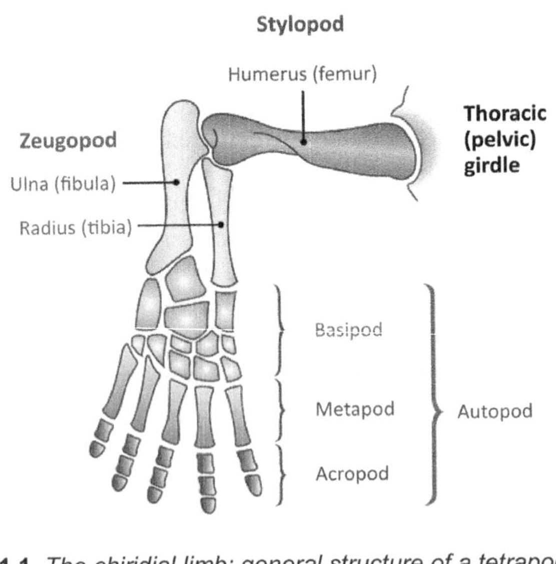

Figure 1.1. *The chiridial limb: general structure of a tetrapod's limb. For a color version of the figure, see www.iste.co.uk/cazeau/foot.zip*

The typical shape is pentadactyl and almost always has a locomotor function.

When our gait is increased or semi-increased, the foot is plantigrade, with five complete segments numbered from 1 to 5, from the tibial (medial) side to the fibular (lateral) side. In quadrupedal gait, the hands and feet have similar functions. The number of segments is reduced, four in the digitigrade of carnivores and even less in unguligrades.

Let us quickly put the tarsus and the zeugo-autopodal joint into context.

The appearance of the chiridial limb in the Late Devonian period (370 million years ago) was a fundamental acquisition that enabled the transition from aquatic to terrestrial life. Anatomically speaking, it was the transition from an aquatic fin to a limb that enabled the conquest of continents. Seventy million years after its alleged disappearance, the famous missing link, the coelacanth fish, was caught in 1938 in the Comoros. Its fin seemed to have the anatomical shape of a section with the chiridial limb.

The structure is as follows (Figure 1.1): the proximal segment or *stylopod* is represented by the arm (humerus) or thigh (femur), the middle segment or *zeugopod* is the forearm (radius and ulna) or leg (tibia and fibula) and the distal segment or *autopod,* consisting of three parts, is the hand or foot. The *basipod* corresponds to the tarsus (and the carpus), the *metapod* to the metatarsus (metacarpus) and the *acropod to* the phalanxes.

Here, we are interested in the ankle (zeugo-autopodial joint). This limb has maintained its general structure but has evolved differently according to lineages.

The ancestors of the first amphibians (labyrinthodontia) were known since the Devonian period. They had an autopod made of multiple parts whose main center of mobility was the joint spaces between the basipod and the metapod (Chopart).

During evolution, numerous fusions took place; a talus and a calcaneus were observed in the whole group at the origin of the mammals, turtles, reptiles and birds (amniotes).

Then, two main developments occurred; the first group came to be represented by turtles (chelonians), most reptiles and birds, and the second by mammals.

In the first, the talus-calcaneus pair is immobilized on the zeugopod, the fibula participates as much as the tibia in the function and the foot's movement is more distal in the tarsus (Figure 1.2).

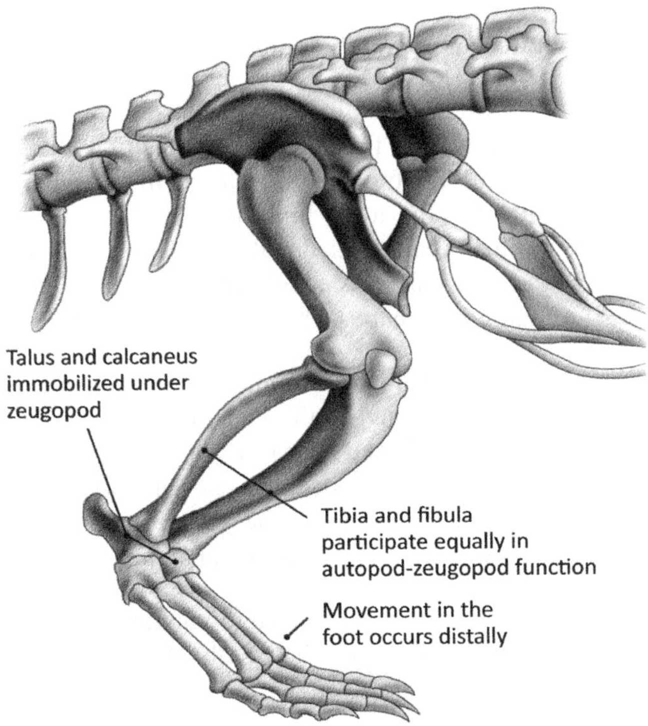

Figure 1.2. *Lower limb of a reptile (crocodile)*

The gastrocnemius muscle remains inserted opposite the plantar fascia (Figure 1.3).

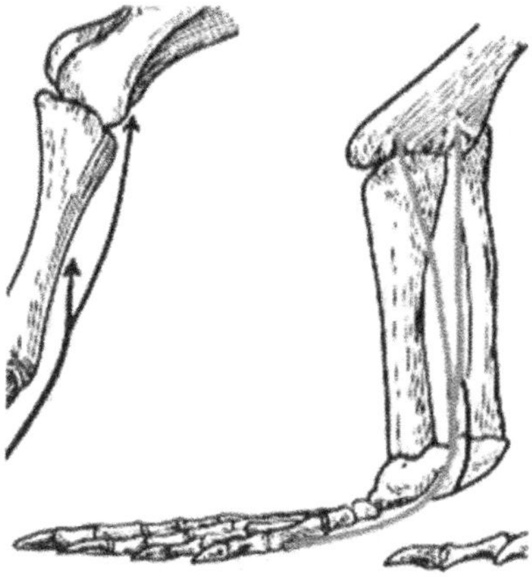

Figure 1.3. *Insertion of gastrocnemius on plantar fascia below mediotarsal joint. For a color version of the figure, see www.iste.co.uk/cazeau/foot.zip*

In mammals, the functional joint is located between the zeugopod and the basipod. The talus is superimposed on the calcaneus which develops the sustentaculum tali, the fibula participates much less in the joint and regresses and the insertion of the gastrocnemius muscle is transferred to the calcaneal tuberosity, effectively increasing its strength (Figure 1.4).

Figure 1.4. *Insertion of gastrocnemius on calcaneus. For a color version of the figure, see www.iste.co.uk/cazeau/foot.zip*

Anatomical adaptations are then made according to the type of locomotion. These concern the entire limb, but the autopod undergoes the most significant transformations (Figure 1.5).

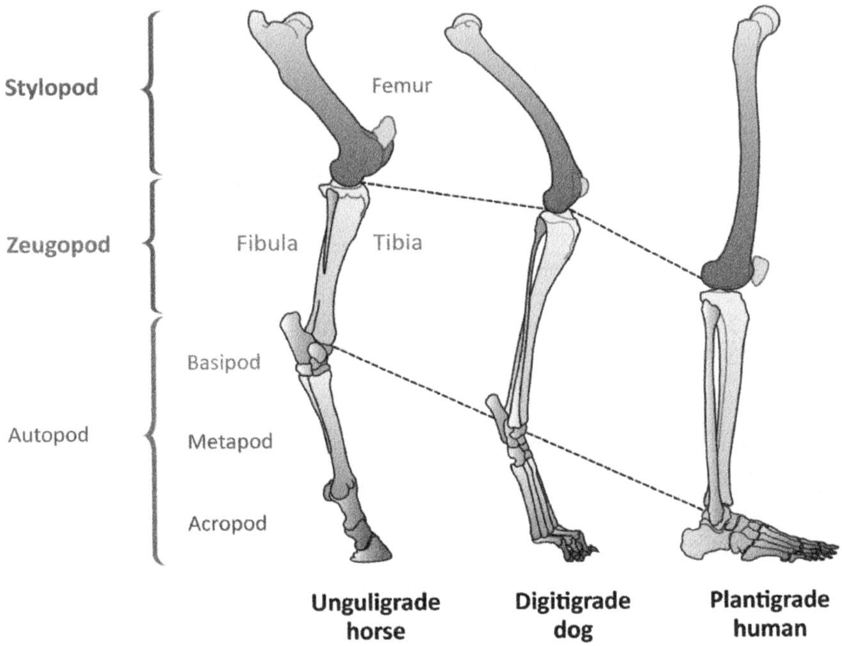

Figure 1.5. *Transformation of lower limb and, in particular, the autopod to match the type of locomotion. For a color version of the figure, see www.iste.co.uk/cazeau/foot.zip*

Adaptation to quadruped running was achieved through three types of simultaneous modifications: lengthening of the limbs (especially at the expense of the metapod), raising of the autopod and progressive reduction in the number of fingers. The same applies to the adaptations for jumping, tree climbing, flying or when returning to aquatic life.

1.2.2. *The human model*

Following this general description, we will focus on the human model from now on.

1.2.2.1. *Bone factors for transverse ankle stabilization*

The transverse stabilizing factors of the ankle concern the bones and are related to the following:

– shape of the talus: extended very transversely, the grooves are wide and the pulley (its anatomical shape) is well developed from front to back, giving a wide stance in an erect position whatever the degree of flexion/extension. The tarsus is wide and robust, adapted to the upright position, and the contact surface between the talus and the tibia is proportionally the most important. The bearing surface is estimated at 4.5 to 9 cm^2;

– narrow mortise and tenon fitting (Figures 1.6 and 1.7): this results from the bimalleolar clamp, the two malleoli descending very low in relation to the talus;

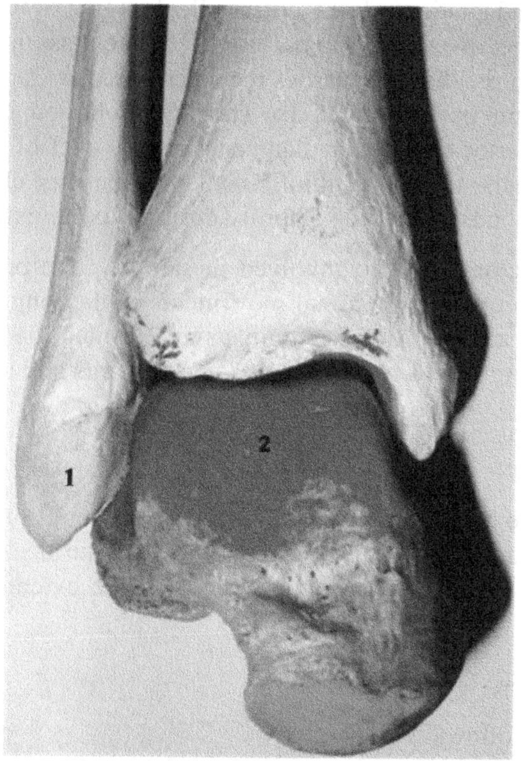

Figure 1.6. *Human crurotarsal joint (front view). 1) Lateral malleolus; 2) tibial articular surface of the talus. For a color version of the figure, see www.iste.co.uk/cazeau/foot.zip*

Figure 1.7. *Proximal side of a human crurotarsal joint (view from below). For a color version of the figure, see www.iste.co.uk/cazeau/foot.zip*

– permanent adaptation of the intermalleolar gap; it varies due to the physiology of the distal tibiofibular joint; in fact, as the talus pulley is narrower backwards than forwards, the intermalleolar gap varies by about 5 mm to maintain perfect transverse stability. Cohesion is ensured by the anterior and posterior tibiofibular ligaments at the level of the syndesmosis. Contrary to animals, the distal tibiofibular joint does not closely connect its two constituent bones, providing stability during flexion/extension;

– ligament factors are also involved in stability; the powerful collateral ligament systems oppose rotational movements in the longitudinal axis. The system, which is already robust in humans and limited to short components, thickens and strengthens from carnivores to ruminants and then to equids.

1.2.2.2. *Principles of anteroposterior stability and factors limiting flexion/extension*

The amplitude of flexion/extension primarily depends on the extent of the proximal articular surface of the talus and the tibial surface. Other factors are bone abutments, ligament tension and, to a lesser extent, the muscle and tendon elements.

1.2.2.2.1. Factors limiting flexion

These are as follows:

– bone factors: in extreme flexion, the upper surface of the talus neck rests against the anterior edge of the tibial pilon;

– capsule and ligament factors: the posterior elements become strained, i.e. the posterior talofibular component, the deep plane of the medial collateral ligament and the posterior capsule, which is flexible.

1.2.2.2.2. Factors limiting extension

These are as follows:

– bone factors: the posterior tubercles of the talus rest against the posterior margin of the tibia, which descends lower than the anterior margin;

– capsule and ligament factors: the anterior elements become strained; anterior talofibular joint, anterior part of the deltoid ligament, anterior capsule, which is flexible;

– muscular factors: in men, the terminal tendons of the muscles in the leg area are short; their tonus is inadvertently significant, limiting joint amplitude, and the possibilities of storing some elastic energy. The muscle mass is low, favoring the erect position but not the lower limb's swing speed. The amplitude of extension is important, as are all the articular segments of the hind limb.

1.2.3. *Comparative functional anatomy*

In the human model, the anatomy of the tibiotarsal joint is very different from that of other mammals, but the result is identical: exclusive flexion/extension.

Transverse stability is ensured by the dynamic quality of the bimalleolar tibiofibular syndesmosis and the short collateral ligaments.

The absence of long components, associated with the shallow tibiotalar joint grooves, has the disadvantage of creating a weakness in this joint, which is partly responsible for the high frequency of ankle sprains.

On the other hand, the absence of these long components has the advantage of not limiting the movements of the midtarsal joint, allowing the inversion of the foot, a condition of the arboreal life of primates, in particular those whose weight is carried by their hind limbs (lemurs and tarsiers).

1.3. Different functional anatomy models of bipedalism, from turkeys to humans and great apes

1.3.1. *Challenges in terms of phylogeny*

Bipedalism is a functional anatomy entity that includes a wide variety of biomechanical mechanisms.

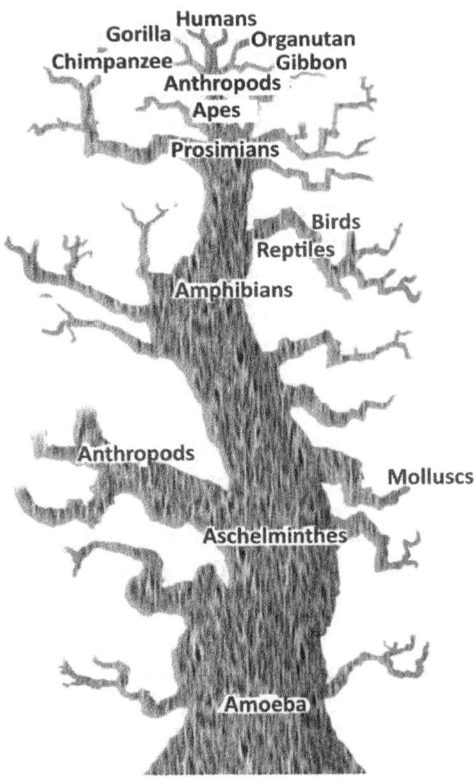

Figure 1.8. *Haeckel's phylogenetic tree placing humans at the top. For a color version of the figure, see www.iste.co.uk/cazeau/foot.zip*

This concept is highly controversial and has been overvalued over the centuries, out of ignorance, belief in a divine creation or to feed the various theories of evolution, with an inseparable anthropocentric feeling of superiority, with humans at the top (Figure 1.8). It is still considered today as one of the answers to the hominization process. This may have led us to forcibly unite in a direct and vertical phylogenetic relationship between apes and humans and to automatically place Australopithecus as an intermediate form.

Yet the human species moves according to a highly specialized, unique mode of bipedalism, very different even from that adopted by animals known phylogenetically to be the closest, i.e. the great apes, of which the bonobo is one.

The choice of a resemblance criterion is as crucial as it is arbitrary, having a direct influence on classification and phylogeny. Thus, the overrepresentation of a morphological or functional feature relative to the complete characteristics of an individual leads to erroneous conclusions.

The "bipedalism" criterion, historically overvalued, thus groups together various entities. Humans do not have exclusivity; the types vary greatly within the different animal classes – present and extinct – and notable differences can be observed among fossil hominoidea, primates and even within the genus *Homo*.

1.3.2. *Definition of bipedalism*

Different types of terrestrial progressions exist: quadrupedalism, bipedalism and crawling aptodes. What does "bipedalism" mean? According to the Latin etymology, it means walking on two feet. According to Sabine Renous, bipedal progression is characterized by the following elements: the vertebrate's body rests entirely on repetitive synchronous or alternating movements of the hind limbs. The forelimbs, freed from being required for terrestrial locomotion, are entirely available for new activities, such as flight in birds or the non-specialized use of the hand in humans.

In the animal world, "bipedalism" generally characterizes movement on the extremities of the lower limbs. This does not presume the support, which can be terrestrial or arboreal only, or mixed. Moreover, this bipedalism can be exclusive as in humans or be integrated into a locomotor repertoire more or less varied in terms of quadrupedalism or the arboreal substrate.

Humans do not therefore have the exclusivity of bipedalism based on these definitions, and anatomy differs according to zoological groups: dinosaurs, birds, primates, bears, etc.

1.3.3. *Different bipeds*

1.3.3.1. *Dinosaurs*

The first known forms to adopt bipedalism were some archosaur dinosaurs, which include most of the reptilian forms of the secondary era (250 to 64 million years ago) (Figure 1.9a and 1.9b), especially during the Jurassic and Cretaceous periods. Authors emphasize the anatomical particularities of the locomotor system. The classical arrangement of the reptilian limbs was then the transverse limb, still observed today in crocodiles (Figure 1.10), barely allowing the body to be lifted. The disadvantage is that considerable forces were required on the limbs to hold the trunk and belly above the ground, causing, among other things, breathing problems due to crushing of the rib cage.

a) b)

Figure 1.9. *Bipedal dinosaurs of the secondary era*

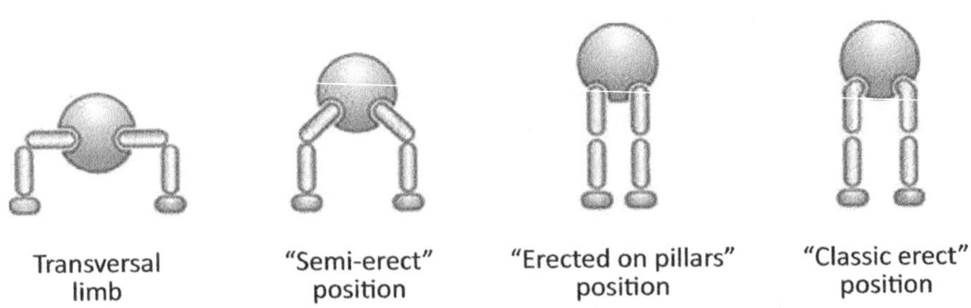

Figure 1.10. *Transition from transverse to parasagittal limb, various types of erect positions*

Based on this model, many archosaurs adopted a semi-vertical bipedal position, with tail development and straightening from the hind limbs. In the middle of the Triassic period, archosaurs experienced a great evolutionary form of disappearance. The evolution was achieved by adopting certain lineages of an erect posture, necessary for bipedalism. In animals that have an erect position, the lower limbs were positioned vertically under the body, the weight transmitted directly to the ground via the limbs, without exerting any leverage (minimum muscular and articular moments). In order for this change to occur, many bone and joint changes happened. Different mechanical models existed by zoologic groups and periods. At the hips, some animals (*Euparkeria*) were in a "semi-erect" position (Figure 1.10), with the acetabulum looking down and out, articulating on a femur that did not have a cervical portion. Others (*Saurosuchus*) were in the "erect on pillars" position, with the lower limbs upright and the femur without a cervical portion fitting into a downward-facing acetabulum. This likely limited the stride range and decreased the force on the hip abductor muscles. Finally, other groups adopted the "classic erect" position, with the acetabulum facing down and out, articulating with a vertical femur but off-center due to the presence of a cervical portion, resembling the current quadruped model. Anatomically, the pelvis, which supports the weight of the abdomen, developed and took on a triradiate shape. The ilium in particular developed and joined – in addition to the sacral vertebrae – with the presacral, lumbar and caudal postsacral vertebrae to form a synsacrum as in birds (Figure 1.11). The femoral heads became stronger and formed a bearing surface, with four bony processes on the femur providing numerous muscle insertions between the pelvis and the femur, as well as and especially between the tail and the femur. The tibia became thicker and the fibula atrophied.

As far as the ankle and foot are concerned, bipedalism was possible in these different groups based on different mechanical models. It may have been a "primitive mesotarsal" ankle with the characteristics of the reptilian intratarsal joint that later came to be found in birds; the distal part of the tibia merging with the posterior tarsus, the anterior tarsus with the metatarsals (Figure 1.12) and the talocalcaneal block being fused to the tibia. Mobility was present in the Chopart and Lisfranc joints. In other groups of bipedal dinosaurs, the mobile hinge was located between the talus and calcaneus (an evolved mesotarsal ankle). In others, it was located in the tibiotalar joint, identical to that of mammals.

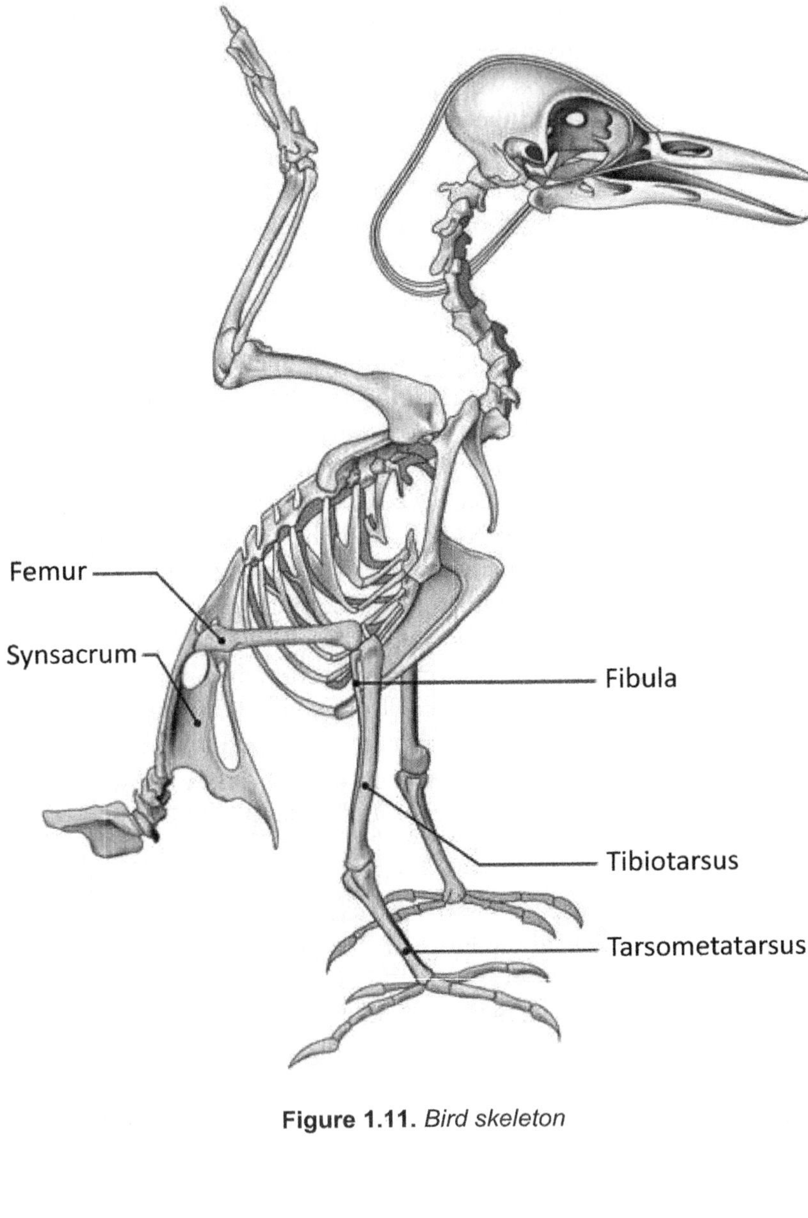

Figure 1.11. *Bird skeleton*

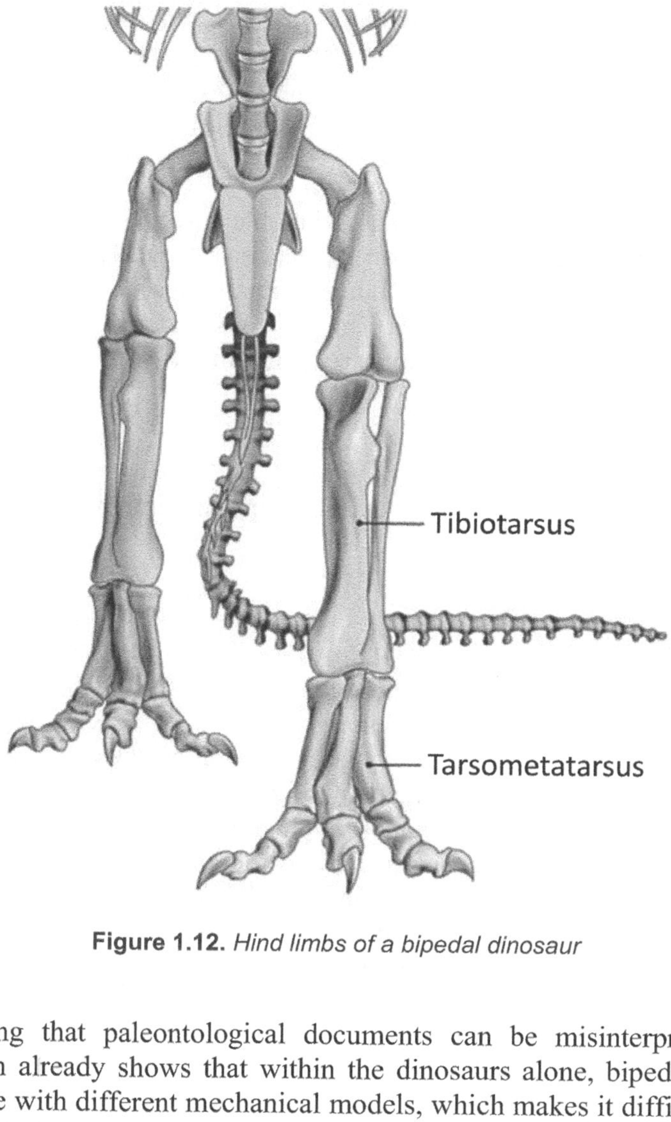

Figure 1.12. *Hind limbs of a bipedal dinosaur*

Knowing that paleontological documents can be misinterpreted, this description already shows that within the dinosaurs alone, bipedalism was compatible with different mechanical models, which makes it difficult to the use this criterion to judge their phylogeny.

1.3.3.2. *Birds*

The second group consists of birds, all bipedal, especially those that do not fly: ostriches, rheas, emus, cassowaries and kiwis. The bipedal bird is of the digitigrade type; its pelvis is triradiated with a synsacrum (Figure 1.11) to support the weight of the body, but with the pubis and ischium reduced to elongated rods, parallel to the spine, probably to enable the delivery of large eggs. The femur is short, always shorter than the tibia, with the iliofemoral angle closed.

Like in archosaurs, a close relative, the tibia is solid, the fibula small and the tarsus partly fused to the tibia and metatarsals, to make up the tibiotarsus proximally (homologous to the human tibia) and the tarsus–metatarsus distally (homologous to the foot). The foot is supported by the phalanxes, with the fifth ray having disappeared and the first ray being directed backwards (Figure 1.11).

1.3.3.3. *Primates*

The third major group is that of primates, the current classification of which is shown in Figure 1.13.

1.3.3.3.1. Human model

In the modern human model (*Homo sapiens*) of permanent terrestrial bipedalism, the wings of ilium are enlarged with shortening of the pubic symphysis, constituting a "pelvis" in the literal sense. The lower limbs have gone from the transverse stage to the erect parasagittal stage and are very mobile in the anteroposterior plane. The femur is significantly elongated, accompanied by eccentricity and strengthening of the femoral neck. The tibia is also attached to the fibula, with no rotation in the knee and ankle, a slight inversion of the foot, and a plantigrade position with rigid longitudinal and transverse foot arches. Integration of the base of the second metatarsal in the midtarsal joint is the most important in all primates, thereby locking it and opposing the midtarsal break found in great apes.

However, even within the genus *Homo*, things are not that simple. Let us take the example of *Homo naledi*, a recent discovery, of unknown age but estimated to date back between 2 and 2.5 million years (Ma). They were essentially human, with a first modern ray, as well as the tarsus and the ankle. However, their feet differed from modern humans with curved phalanxes and a reduced longitudinal arch.

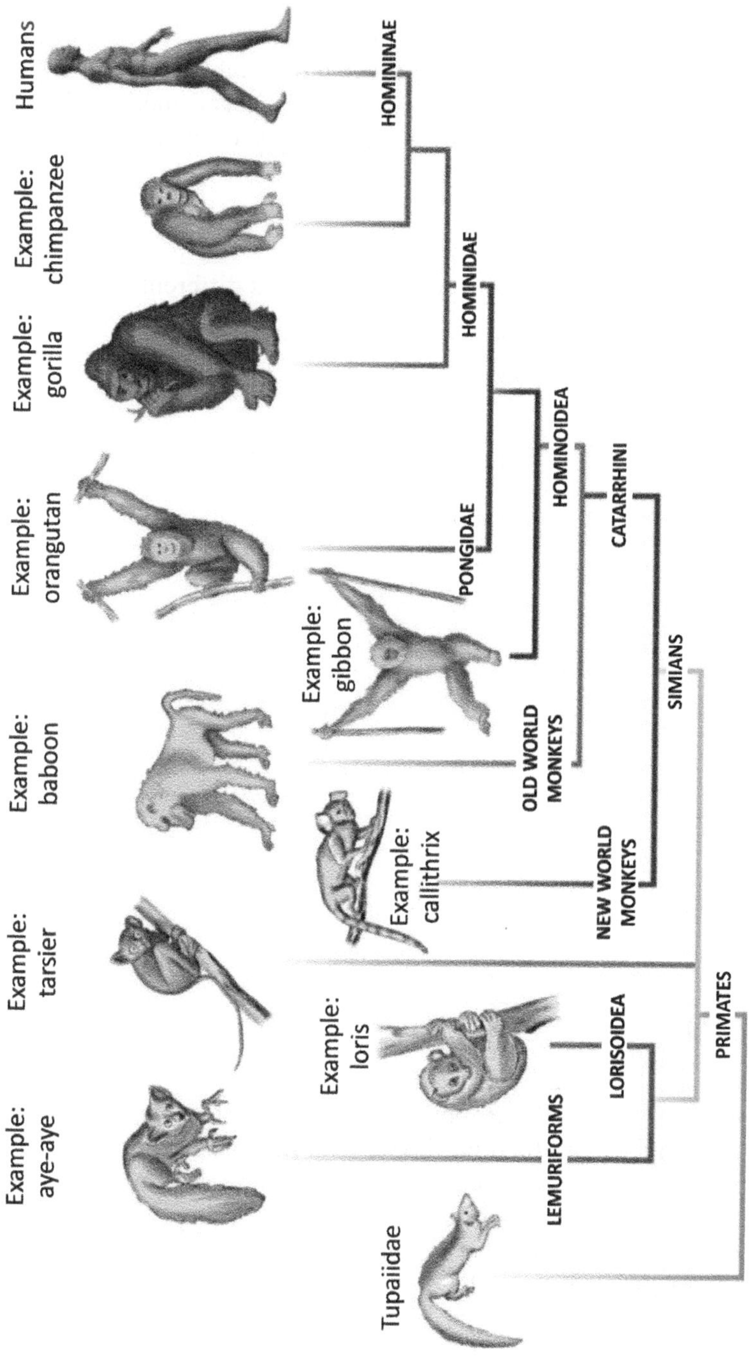

Figure 1.13. *Classification of primates. For a color version of the figure, see www.iste.co.uk/cazeau/foot.zip*

1.3.3.3.2. Australopithecus

In the Australopithecus, which are older terrestrial bipeds (mostly dated between 5 and 3 Ma), the mixture between archaic and modern features is even more marked. Even in these ancient forms of permanent terrestrial bipedalism, there was great locomotor diversity.

1.3.3.3.3. Great apes

In great apes, such as chimpanzees, the pattern is different.

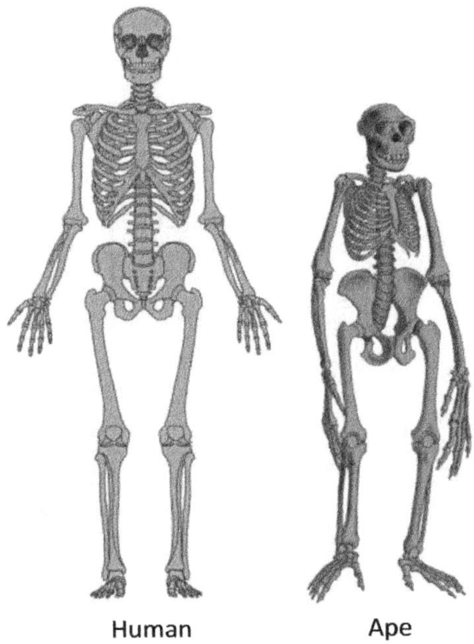

Figure 1.14. *Human and chimpanzee skeletons, front view*

The general posture is flexed hips and knees (Figures 1.14 and 1.15). The forelimbs are longer than the hind limbs, with the chest more of an inverted funnel shape. The spine has only two curves, not very marked, and the lumbosacral angle is very significant. The pelvis is high and narrow with flat, open wings of ilium, holding the viscera less than the human pelvis. The femurs are vertical, with round condyles at the knee that allow rotation.

As for the ankle and foot, we use the elements described by Yvette Deloison. From the proximal segment, the differences are notable: the talar joint surface is quadrangular in humans, allowing only flexion/extension, while the triangular surface of the other great apes allows foot rotation. The human calcaneus, which is angled upwards and forwards, is responsible for initiating the longitudinal arch of the foot and rests on the ground on a lateral and medial tuberosity ensuring its stability. The opposite is true in other species, characterized by a horizontal calcaneus with a profile resting frontally on its single medial tubercle, thus arranged in a varus shape.

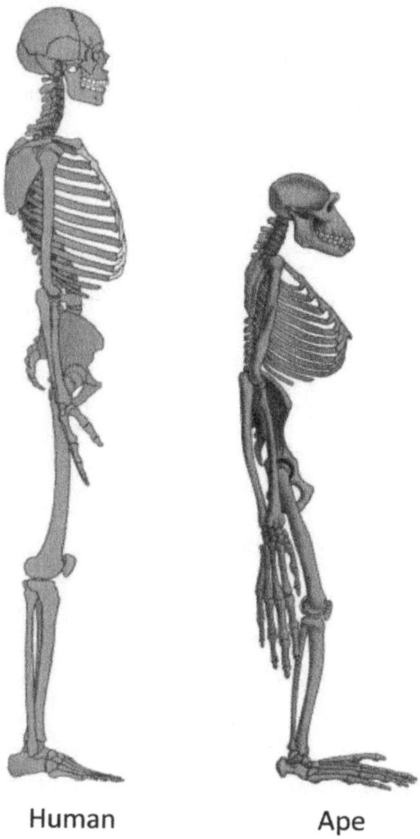

Figure 1.15. *Human and chimpanzee skeletons, side by side*

The proportions are also different. The size of the foot relative to the hind limb is smaller in humans, but the tarsus is proportionally larger

(Figures 1.16 and 1.17). The long bones are rather straight with triangular cross-sections in humans, whereas in the other apes, the bones are strongly concave in the plantar region, with rounded cross-sections, on a proportionally very long forefoot. The shape of the metatarsal paddle is not greatly different, unlike the first ray, but nevertheless the lateral rays are twisted into inversion, placing the heads opposite that of the hallux, thus favoring the gripping function and arboreal locomotion (Figure 1.18). In humans, the first metatarsal is almost parallel (10° of divergence) to the other four, not very mobile, connected by an arthrodia to the distal face of the first cuneiform, and prolonged by a hallux more or less in valgus and not opposable to the others. The opposite is seen in the other hominoids; the first metatarsal is connected to its cuneiform by a trochoid-shaped joint allowing rotation, opposition and opening of the first web space, responsible for the varus position.

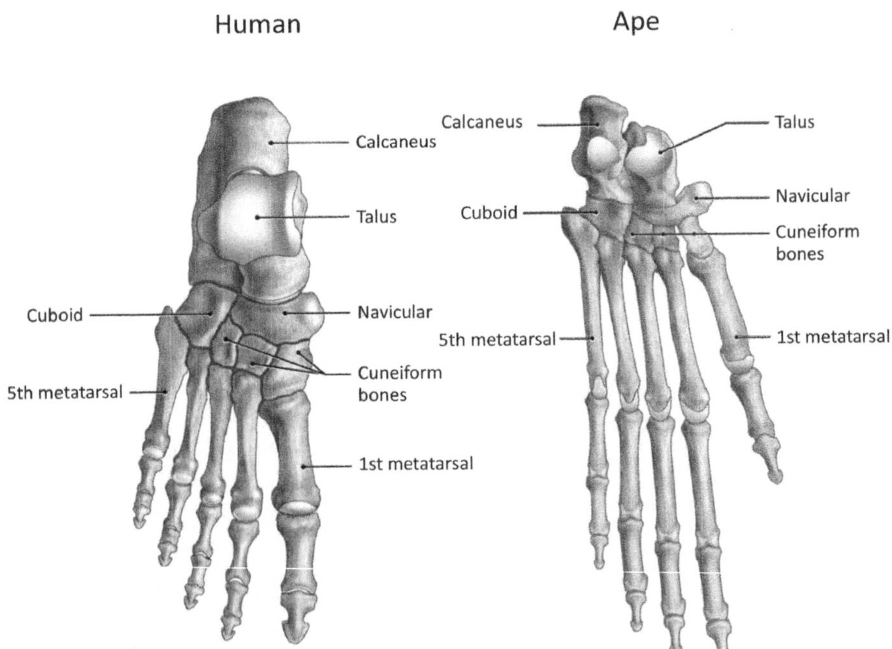

Figure 1.16. *Skeletal structure of a human and chimpanzee foot, front view*

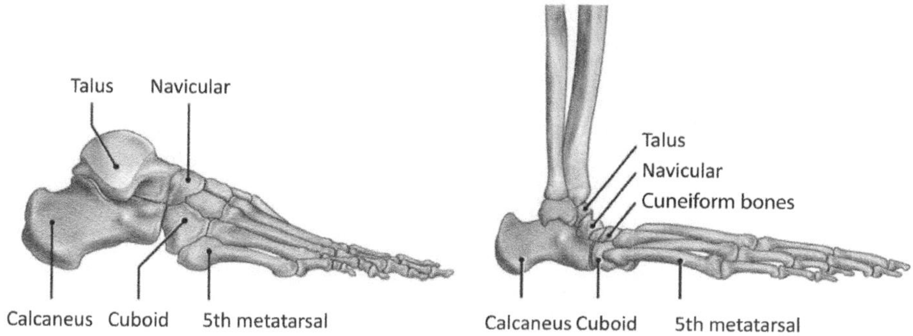

Figure 1.17. *Skeletal structure of a human and chimpanzee foot, lateral view*

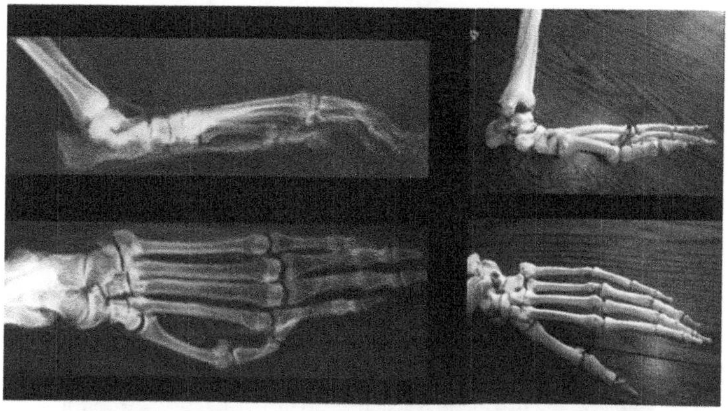

Figure 1.18. *Monkey foot, radiographs and skeleton*

With regard to muscles (Figures 1.19 and 1.20), the major difference is the considerable amount of muscle mass used to mobilize the hallux. Simians in particular, unlike humans, have a very powerful short abductor, short flexor and adductor muscles and even a long abductor, which is absent in humans. This makes the first ray highly mobile, thus very useful in arboreal locomotion. The distal insertions of the long flexor are very different; their intermeshing on the plantar face provides active muscular structural cohesion, which is opposed to the passive cohesion of the human plantar arch provided by the bone and ligament structures.

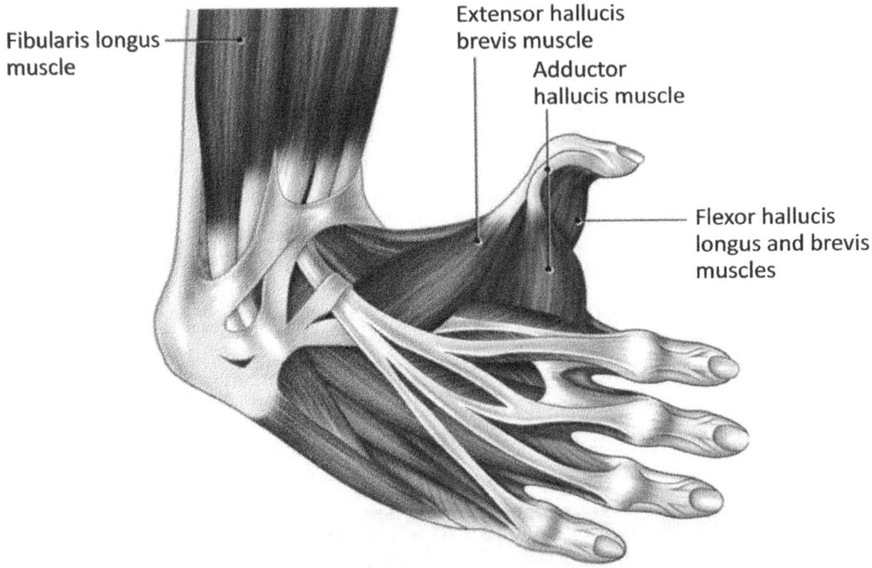

Figure 1.19. *Chimpanzee foot muscle. For a color version of the figure, see www.iste.co.uk/cazeau/foot.zip*

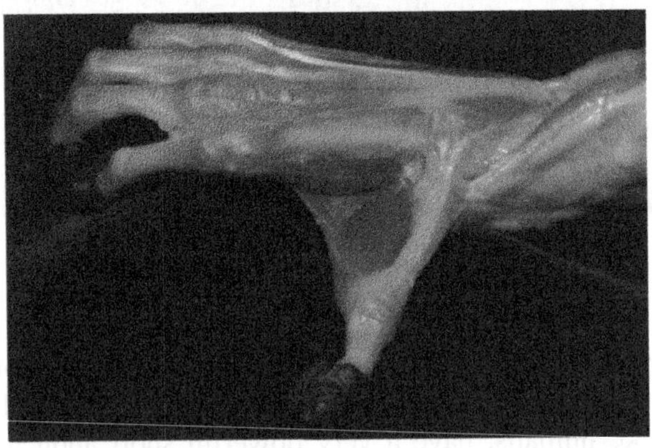

Figure 1.20. *Baboon foot. Note the large amount of muscle mass dedicated to the hallux (dissection of a baboon – Pau Golano, anatomist; Cyrille Cazeau, surgeon; Gilles Berillon, anthropologist). For a color version of the figure, see www.iste.co.uk/cazeau/foot.zip*

Other animals such as the bear use bipedalism, which has an even more different pattern.

1.3.4. *Conclusion*

The "bipedalism" analysis criterion is certainly useful provided that we are talking about the model of permanent modern terrestrial bipedalism. It is specific in terms of anatomy, function and biomechanics. Its analysis is as important for the question of the origins of humanity as it is for its implications in surgical choices. This high specificity leads us to reflect on its origin, taking into account the irreversible laws of evolution.

However, the confusing use of this entity can lead us to suggest that phylogenetic relationships with other great primates are direct by filiation and to automatically place fossil hominoids (Australopithecus) as intermediaries.

1.4. Permanent terrestrial human bipedalism, explained through the model of the bi-articular muscles

The kinematics of the knee and ankle have direct consequences on the functioning of bi-articular muscles of the crural and leg sectors. This unique feature is specific to permanent human terrestrial bipedalism. This is the subject of a mechanical study, with its kinetic, dynamic and electromyographic aspects.

1.4.1. *Introduction*

The specificity of human bipedalism is due to the economy of bi-articular muscles, allowing the storage and restitution of elastic energy.

We are interested here in the gastrocnemius muscle (GM), whose effectiveness is linked to simultaneous extension of the knee and dorsal flexion of the ankle.

We propose to study its functional anatomy through dynamics, kinematics and electromyography. Particular attention will be paid to extension of the knee and dorsal flexion of the ankle when walking, an original feature that is absent even in great apes.

The GM muscle has two femoral heads. They meet on the median line and join with the tendon of the soleus muscle to form the calcaneal tendon.

The GM is a bi-articular muscle, crossing the knee and ankle.

The accepted representation of a muscle on the mechanical and physiological level is the Hill model with three components: a contractile component (actin–myosin bridges), a "stiff" serial elastic component (actin–myosin bridges + calcaneal tendon), which cannot easily be distended, and a "compliant" parallel elastic component (the muscle's connective tissue), which can more easily be distended.

Physiological studies show that *when an already stretched muscle is activated, the force it develops is increased by prior elastic energy storage.* The work in shortening is more significant, and the output is increased. This is the case of a bi-articular muscle, which is necessarily stretched when executing a movement according to the principle of "storage-restitution" of elastic energy described by Cavagna in 1968.

Furthermore, the calculation of forces and moments according to Newton shows that a bi-articular muscle allows the transfer of mechanical energy from one link to the other, further minimizing losses.

1.4.2. *Kinematic analysis*

Kinematics is the study of the effect of forces, such as displacement, speed and acceleration.

To illustrate the GM functional anatomy, we are interested in the walking phase in which its brevity could be revealed, which is the stance phase. Indeed, the oscillation swing phase is characterized by the combined flexion of the knee and ankle in the plantar position, a situation of maximum GM relaxation, which does not unmask its possible brevity.

Changes in hip, knee and ankle angles during the stance and swing phase are illustrated in Figures 1.21 and 2.34.

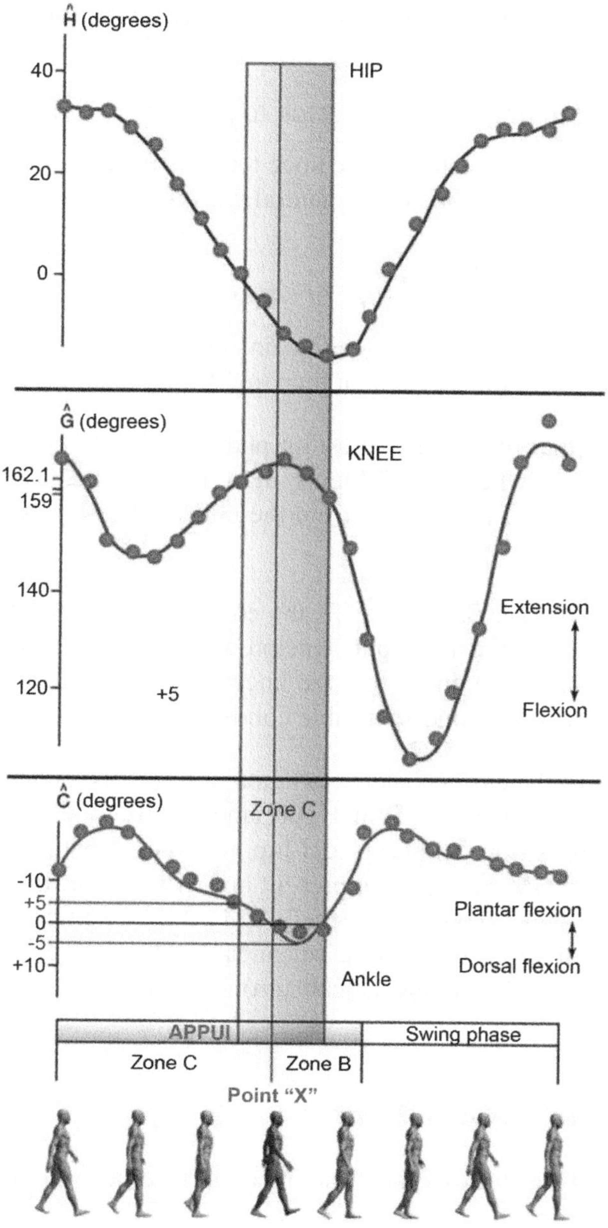

Figure 1.21. *Kinematics of the gait cycle in humans . For a color version of the figure, see www.iste.co.uk/cazeau/foot.zip*

Changes in the hip, knee and ankle angles are shown during the bipedal gait cycle.

1.4.2.1. *From the beginning of stance to point X*

The hip enters into extension, the knee remains in extension and the ankle passes from plantar flexion to the neutral position represented by point X (0°).

1.4.2.2. *From point X to the start of walking*

The hip is in extension, the knee remains close to extension and the ankle goes from 0° to 5° of dorsal flexion, then to 20° of plantar flexion.

We can see that the most interesting phase is shortly before and shortly after the moment when the ankle is at a right angle. This moment is identified by point X, which goes from the 60th to the 88th percentile of the stance phase.

Indeed, in what we call "zone C", the combined effect of knee extension and dorsal flexion of the ankle puts tension on the GM, due to its bi-articular character. It is only at point X that the knee is in extension and the ankle in dorsal flexion, which are indispensable conditions for stretching the GM.

The great ape, the closest relative to humans, has in its locomotor repertoire terrestrial bipedalism, but it is not permanent thus totally different in its design. Walking is done in permanent hip and knee flexion. Thus, although anatomically the GM muscles are present and bi-articular, they cannot tense themselves and make the system benefit from the properties described. This is all the more true since it has no calcaneal tendon (direct insertion of the gastrocnemius muscle on the calcaneum), and therefore no serial elastic component.

1.4.3. *Dynamic analysis (Figure 1.22)*

Dynamics is the study of forces acting on a material system.

A force platform records the external forces transmitted by the body. These forces are made up of the weight and the contact actions between the ground and the plantar surface.

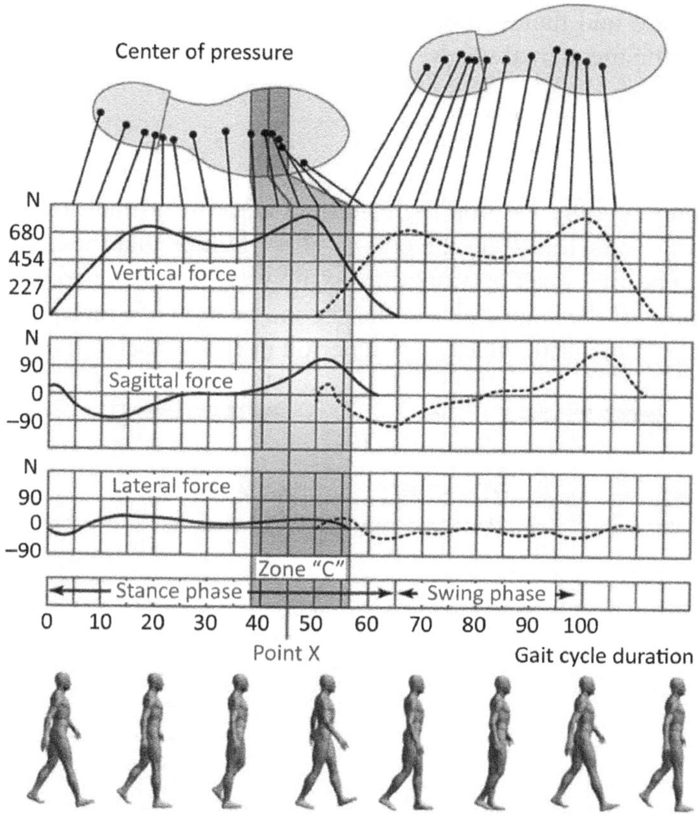

Figure 1.22. *Dynamics of the gait cycle in humans*

They are broken down into an anteroposterior sagittal force representing propulsion, a vertical force responsible for the raising/lowering of the center of gravity and a lateral force in the frontal plane. This allows us to calculate the acceleration, velocity and displacement of the center of gravity.

It is also possible to calculate the center of pressure, which can be used to determine the trajectory of the foot on the ground. This device also provides information on the pressure distribution under the foot when walking.

There is a rise in the center of gravity in zone C as the hip passes vertically over the ankle, followed by a fall, reflecting a decrease in vertical force, with an increase in forward sagittal force, reflecting loss of balance by falling forward and then catching up. The center of pressure in zone C is at the level of the metatarsal heads.

1.4.4. *Electromyography*

Electromyography records the electrical activity in muscles, such as the GM (Figure 1.23). Based on the curves, it is obvious that the activity comes in surges. The first peak appears during the weight transfer to the ground at the beginning of the stance phase; the second peak is of greater interest to us because it occurs during phase C, the period of limb acceleration.

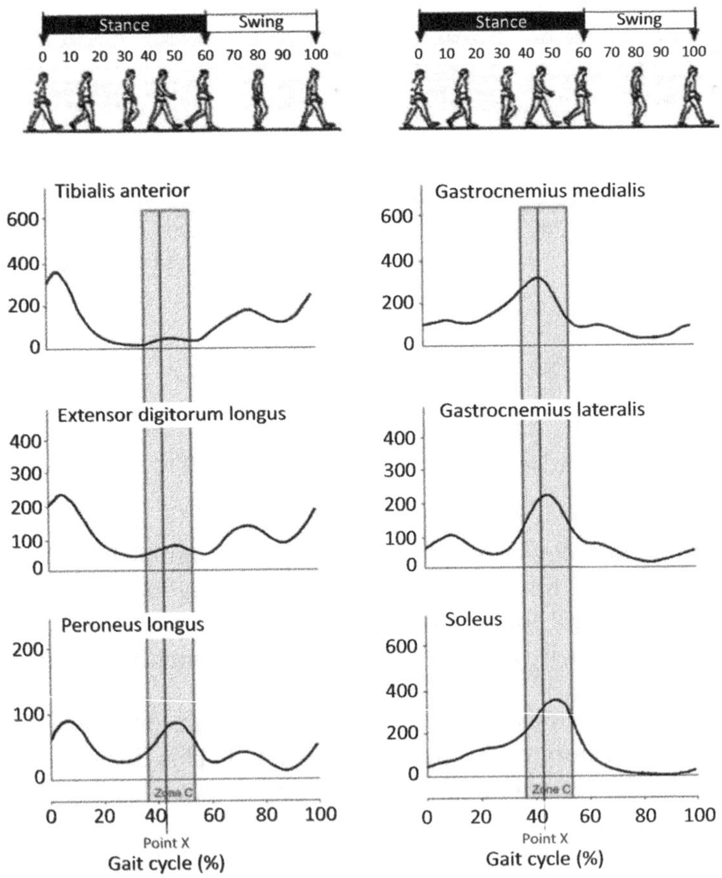

Figure 1.23. *Electromyographic analysis of the gastrocnemius*

The second surge is integrated into the function of the bi-articular muscle.

In Figure 1.24, the curves are superimposed; the angular displacement of the ankle is in black, the EMG of the GM in green, with the exact moment of triggering its second surge of activity occurring at point X. The ankle is at 0°, then enters into dorsal flexion. The knee is in maximum extension.

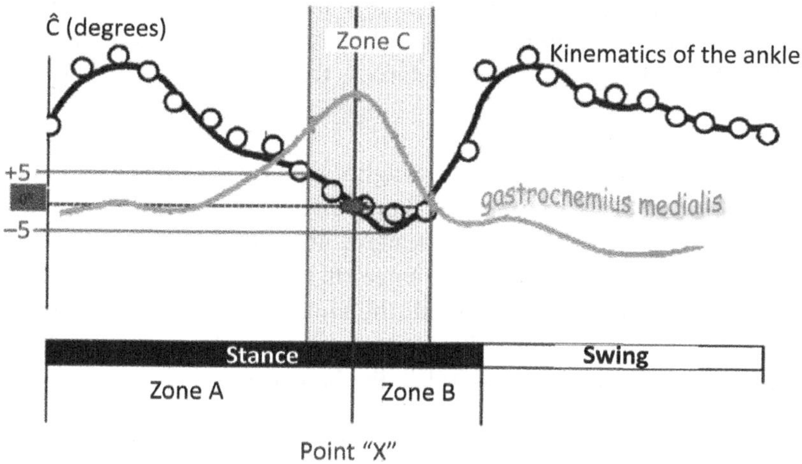

Figure 1.24. *Superimposition of the gastrocnemius muscle EMG and ankle kinematics curves*

1.4.4.1. *Start of stance phase*

At the beginning of the stance phase, there is ankle plantar flexion that is not related to muscle activity but to knee extension that stretches the GM.

1.4.4.2. *At point X*

At point X, the knee is in maximum extension and the ankle is at 0°. The GM is in near-maximum tension. The center of pressure is under the head of the metatarsals. The stored mechanical energy is at its maximum. The EMG spike occurs at this time, considerably increasing the effectiveness of the contraction.

1.4.4.3. *End of phase C after point X*

From the end of phase C after point X, the EMG surge ceases early enough to allow dorsal ankle flexion, which then helps knee flexion in the next phase.

Moreover, dorsal flexion of the ankle is not related to activation of the antagonistic muscles as shown by the low electrical activity of the *tibialis anterior* and *extensor digitorum longus* (Figure 1.23). It is the forward fall of the body (Figure 1.22), i.e. external forces, that passively causes this dorsal flexion.

It is interesting to know that throughout phase C when the knee is in extension, the EMG activity of the knee extensors (*rectus femoris*, *vastus lateralis*, etc.) (Figure 1.25) is at its lowest, illustrating the economy of these other bi-articular muscles above the knee. It is the forward fall of the body, with the foot fixed on the ground and the ankle in maximum dorsal flexion, that causes the knee to extend. This in turn provokes hip extension by means of a third group of bi-articular muscles, the hamstrings.

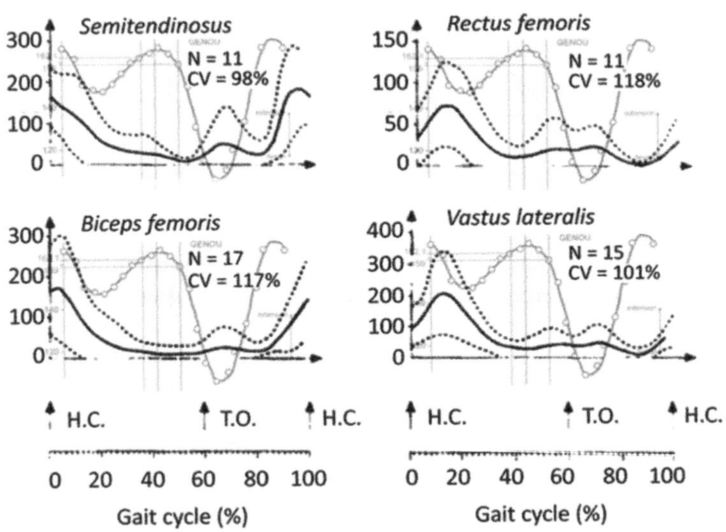

Figure 1.25. *EMG; analysis of peak leg muscle activity during the bipedal gait cycle*

This is a good illustration of the intertwining of electrical muscle activities with the functioning of the bi-articular systems, during the execution of coordinated movements, here characteristic of human walking.

1.4.4.4. *End of stance phase and beginning of swing phase*

Dorsal flexion of the ankle on a knee still in extension from the previous phase stretches the GM, which then causes the knee to flex, as well as the passage of the ankle into plantar flexion, not linked to the EMG activity. It is only afterwards that the EMG activity of the knee flexors (sartorius, biceps femoris, semitendinosus) increases during the swing phase (Figure 1.26).

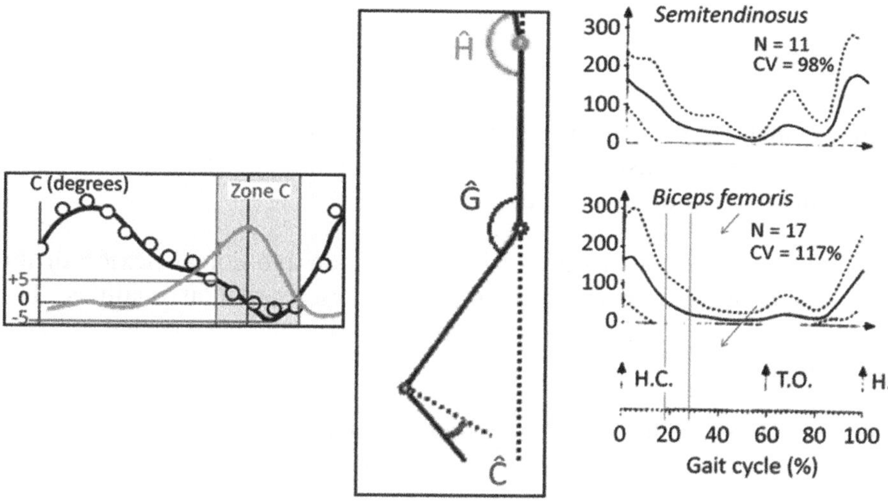

Figure 1.26. *EMG analysis of peak knee extensor activity over the bipedal gait cycle*

1.4.5. *Surgical treatment*

The treatment for short GMs is to make them longer by stretching them. The most reliable technique is removing the medial GM from the popliteal fossa. This is close to the anatomical norm for regaining useful function (Figure 1.27).

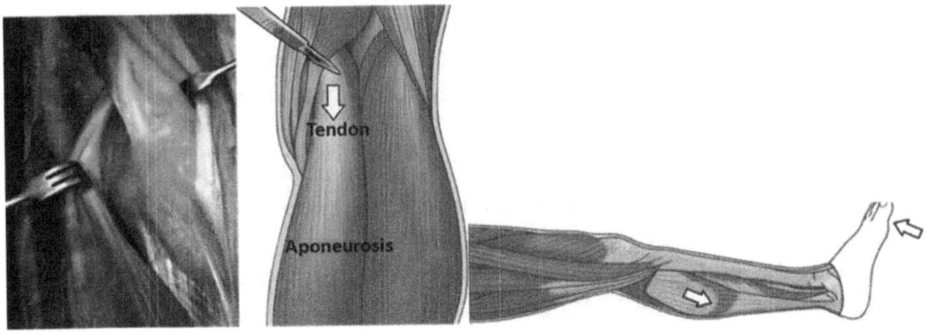

Figure 1.27. *Disinsertion of the medial gastrocnemius (source: Dr Samuel Barouk)*

1.4.6. *Conclusion*

Bi-articular muscles, such as the gastrocnemius used as an example here, help to save and transfer energy from one phase of gait to the other, constituting two major advantages.

It is the simultaneous nature of knee extension and dorsal flexion of the ankle in supporting humans that highlights these particular qualities. The knee, on the other hand, also belongs to the bi-articular system that unites it with the hip.

This highly specialized system is fundamentally different from that of the great apes and should therefore lead us to question our phylogeny.

1.5. Alternating dorsal/plantar flexion of the ankle; comparison with the great ape

The data are taken from Kristiaan's August article that studied the kinematics of bipedal and quadrupedal gait in the bonobo (great ape). The angles have been recalculated to be comparable with those of humans.

Walking is done in permanent hip and knee flexion, with the trunk bent forward (Figure 1.15). The changes in the hip, knee and ankle angles are shown in Figure 1.28 and Table 1.1.

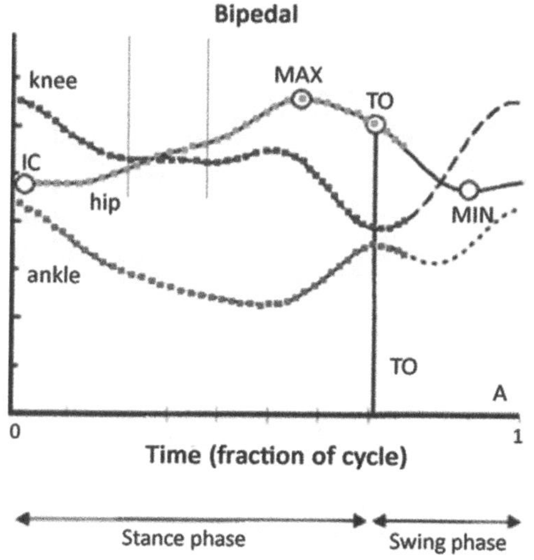

Figure 1.28. *Kinematics of lower limb joints during the stance phase in the bonobo (great ape). For a color version of the figure, see www.iste.co.uk/cazeau/foot.zip*

1.5.1. *Kinematic analysis*

	Initial contact (IC)	Stance phase	Toe off (TO)
Hip	71° Flexion	+30° Flexion	+41° Flexion
Knee	+32° Flexion	+56° Flexion	+93° Flexion
Ankle	–[0–3°] Plantar flexion	+52° Dorsal flexion	+12° Dorsal flexion

Table 1.1. *Hip, knee and ankle angles at three key moments of the stance phase in the bonobo*

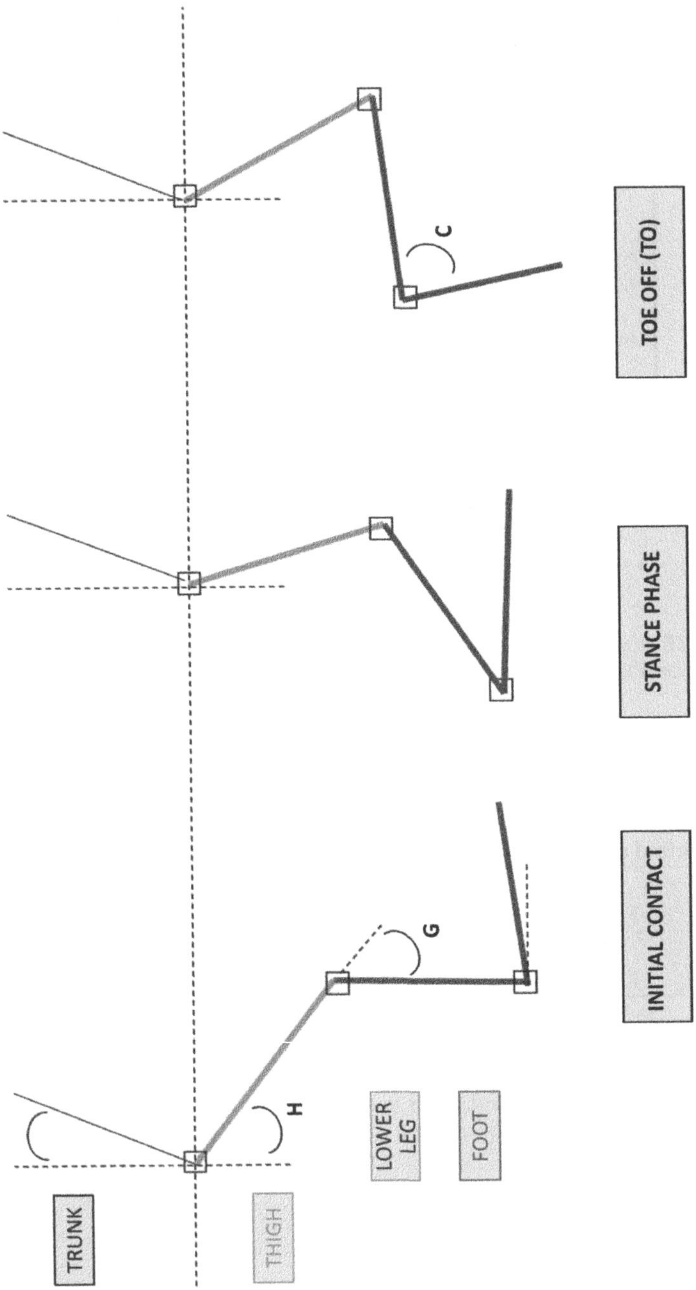

Figure 1.29. *Kinematics modelling of the bonobo stance phase. For a color version of the figure, see www.iste.co.uk/cazeau/foot.zip*

The hip and knee are permanently flexed and can never extend, unlike in humans (Table 1.1).

For the ankle, the beginning of initial contact (IC) is almost at 0°, then dorsal flexion increases during the stance phase up to 52° and returns in the plantar direction without exceeding 12° of dorsal flexion at the time of lift-off. The ankle is therefore in permanent dorsal flexion. In Figure 1.29, we modeled the lower limb as a stage, respecting the angular values in the table. Note that the midtarsal break is not shown in Figure 1.30.

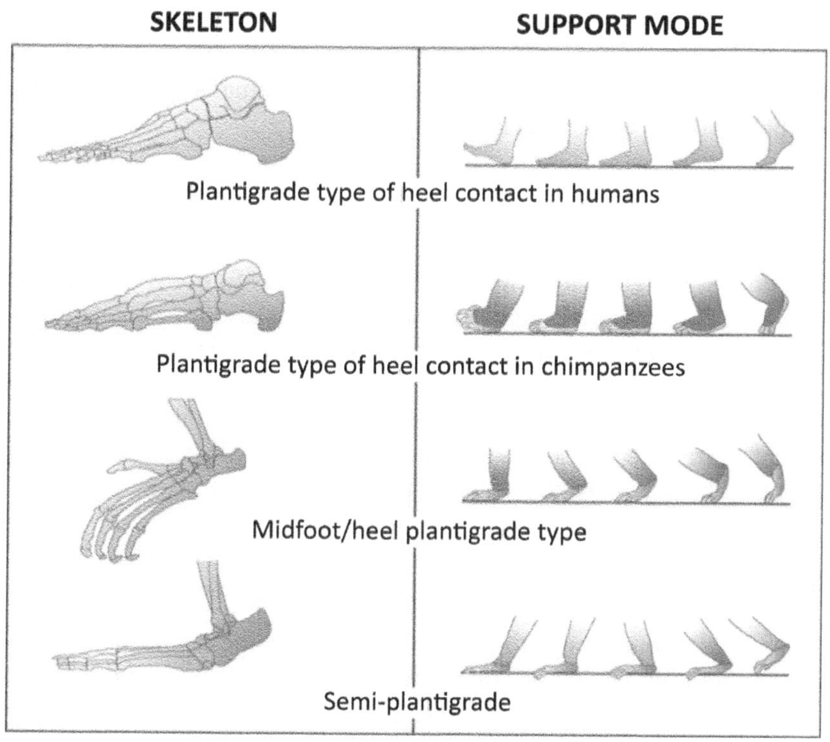

Figure 1.30. *Different modes of stance and locomotion in primates (midtarsal break visible on the semi-plantigrade model)*

We were particularly interested in how the knee/ankle couple functions: although the ankle is in permanent dorsal flexion, knee flexion during the entire stance phase prevents the gastrocnemius muscle from tightening at any time and therefore allows the system to benefit from its bi-articular properties.

The step phase is the equivalent of the C zone defined in humans: it precedes standing up (TO), before forward movement. At this moment, we expect to see the biomechanical parameters promoting forward movement, but we note, on the contrary, that the gastrocnemius is completely relaxed. Indeed, the knee is in maximum flexion and the ankle in minimum dorsal flexion. This illustrates how different the biomechanical mechanisms of bipedalism are.

It is also interesting to note that the coordination plan of the bonobo's bipedalism – which is very different from that of humans – is very similar to that of the bonobo's quadrupedalism.

1.5.2. *Dynamic analysis*

In bonobos, the vertical amplitude of the center of gravity, respecting the scale, is low, due to the fact that the knees and hips are in permanent flexion. The absence of extension minimizes the travel of the center of mass, which is further reduced by the ascent of the calcaneus during walking, responsible for minimizing the functional length of the lower limb. This ascending movement of the heel is linked to the midtarsal characteristic of the great ape's foot mobility, also called the midtarsal break.

1.5.3. *Conclusion*

Under these conditions, the model presented above is vastly different from the human model. There is no efficient transfer of potential energy (rise and fall of center of gravity) into kinetic energy (propulsion), which characterizes the human mechanical model of an inverted pendulum.

As for the apes having permanent flexion of the trunk, hip and ankle, it seems obvious that the energy expenditure in terms of moments is higher than that spent straight, to the detriment of propulsion. The cost of an erect position is higher, and the efficiency is lower.

On the other hand, dorsal flexion of the ankle in humans during the stance phase shows the biomechanical qualities of this economical system. This highly specialized system is fundamentally different from that of great apes and should therefore lead us to question our phylogeny.

1.6. Dorsiflexion location (Figure 1.15)

This is not a *topos* in the Greek rhetorical sense, but functional anatomy. The comparison between apes and humans shows that the two models are opposed despite an apparent similarity. The latter's foot is characterized by the locking of the midtarsal by enclosing the base of the second metatarsal, and the loss of first ray opposition. Gait is postponed distally in terms of the metatarsophalangeal joints (MTPs) dedicated to extension (Figure 1.31).

The opposite is true for apes, whose footsteps pass through the mobile midtarsal joint (midtarsal break). MTPs function mainly in flexion, adapted to gripping and arboreal quadrupedalism (Figure 1.32a and b).

Comparative anatomy allows the specificity of each model to be captured and the treatment to be directed towards restoring function. This will be described in particular in section 2.3 on arthrodesis.

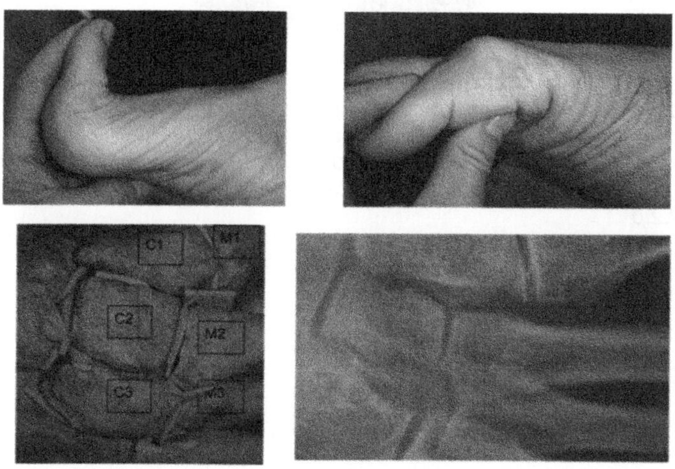

Figure 1.31. *Human gait: locking of the midtarsal area by enclosing the base of the M2 and MTP hyperextension. For a color version of the figure, see www.iste.co.uk/cazeau/foot.zip*

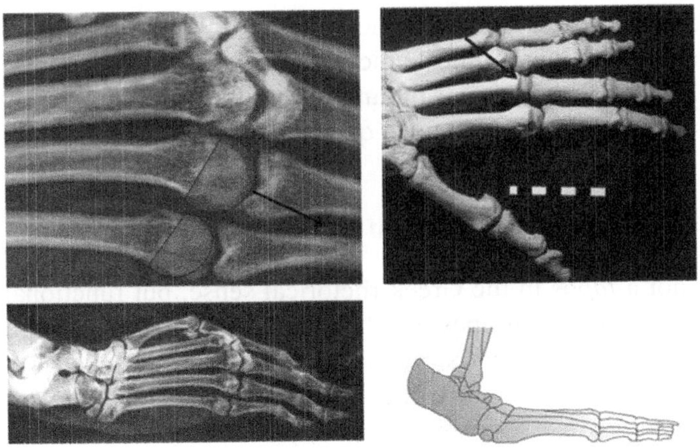

a)

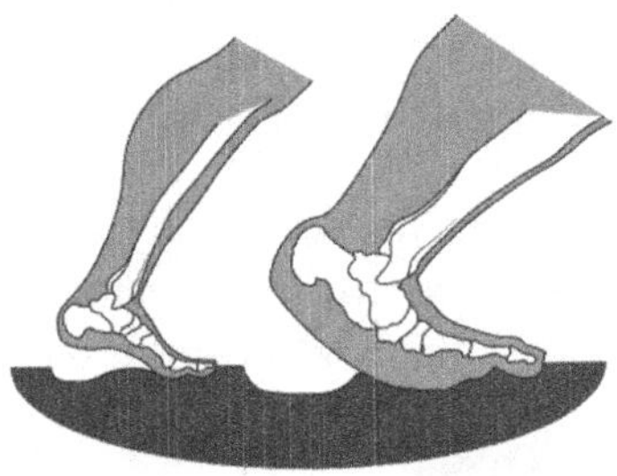

b)

Figure 1.32. *a) Monkey gait: dorsal midtarsal flexion (midtarsal break), the MTP articular facets are oriented in flexion and a retrocapital metatarsal bone bar prevents MTP extension. b) Human gait: distal push-off (MTP) on the left, proximal gorilla push-off on the right (midtarsal). For a color version of the figure, see www.iste.co.uk/cazeau/foot.zip*

2

The Aim of Surgery: Restore Function

2.1. Introduction

2.1.1. *The notion of sacrifice*

When faced with a musculoskeletal system pathology, the possible surgical options are:

– acting on the soft tissues (tendons, joint capsule, etc.), by transecting, lengthening or diverting;

– acting on the bones, by realignment or shortening osteotomies;

– acting on the joints, by resection or arthrodesis.

Nothing is possible without making sacrifices. How then can we choose the most efficient and least aggressive treatment?

The first step is to differentiate and prioritize what separates "normal" from "useful".

2.1.2. *Normal*

Normal has two meanings. The first involves a value judgment: normal is that which conforms to a norm (moral, social, religious, ethical, etc.), to a rule considered valid, legitimate, fair, or to a state considered "healthy". It is the rhetorical equivalent of written law.

The second has no value judgment: normal is what is in line with the average, what is done, happens regularly, is habitual, common. In the context of surgery, we can reason exclusively within the sphere of this dual normality. The anatomical norm with value judgment is the original anatomy, the one described in books, with the exception of anatomical variability. The surgical standard with value judgment is to repair a disorder using a technique that is regularly implemented, disseminated by medical authorities, even if it is static. Thus, in the case of an intra-articular fracture that has seriously damaged the cartilage, the opposite path will be taken surgically, repairing the damaged anatomy as faithfully as possible to restore it to its original anatomy. A third way of thinking is not to define anatomy as a norm but a function – a useful function – to the detriment of anatomical normality.

No one is forcing a debate on standards. However, Yves Schwartz (Schwartz 2014, author's translation) points out that these debates specify our humanity. "The notion of norm has value only in the human kingdom because it is situated in history, it is history". Let us hope the history that surgery constructs can integrate choices of norms that do not leave the desire for omnipotence in fixed medical knowledge such as that of creationism. Without refuting the notion of normal, we can try to emerge from the pre-operative darkness by using very different and independent analysis tools, and not stop at repairing what has been damaged to make it resemble its initial state.

2.1.3. *Useful*

These tools are represented by the comparative functional anatomy approach. Biomechanics, not in a static and unloaded position, but in a supported and dynamic position, initiate the idea "that there are always choices to be made, that since not everything is equal; there are values according to which we will decide". This idea, which is imposing itself as a new norm, involves considering what is useful to the individual in terms of function, no matter if we preserve the original anatomy. In the case of an intra-articular fracture, it might be better to lock the affected joint; an *ad integrum* repair would certainly preserve the original anatomy and look good on X-rays, but would result in disastrous function due to persistent pain.

2.1.4. *Direct surgical implications*

NOTE.–

EDB = *extensor digitorum brevis*

EDL = *extensor digitorum longus*

FDB = *flexor digitorum brevis*

FDL = *flexor digitorum longus*

ADD = *adductor hallucis*

This has direct implications in surgery when making choices. The short muscles of the first ray are quite small in the human foot (Figure 2.1), such as the oblique and transverse components of the adductor (*adductor hallucis*).

1. Flexor halluci brevis
2. Adductor hallucis (oblique bundle)
3. Adductor hallucis (transverse bundle)
4. MP1 lateral collateral ligament
5. Sesamoid phalangeal ligament
6. Sesamoid metatarsal ligament

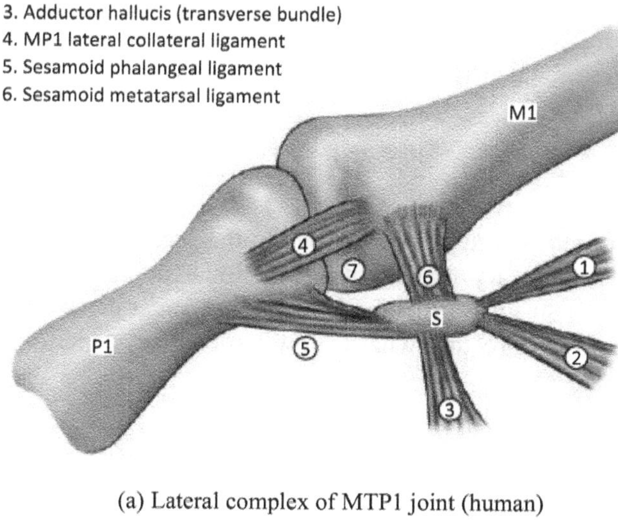

(a) Lateral complex of MTP1 joint (human)

Figure 2.1A. *Short first ray muscles in humans relative to apes. For a color version of the figure, see www.iste.co.uk/cazeau/foot.zip*

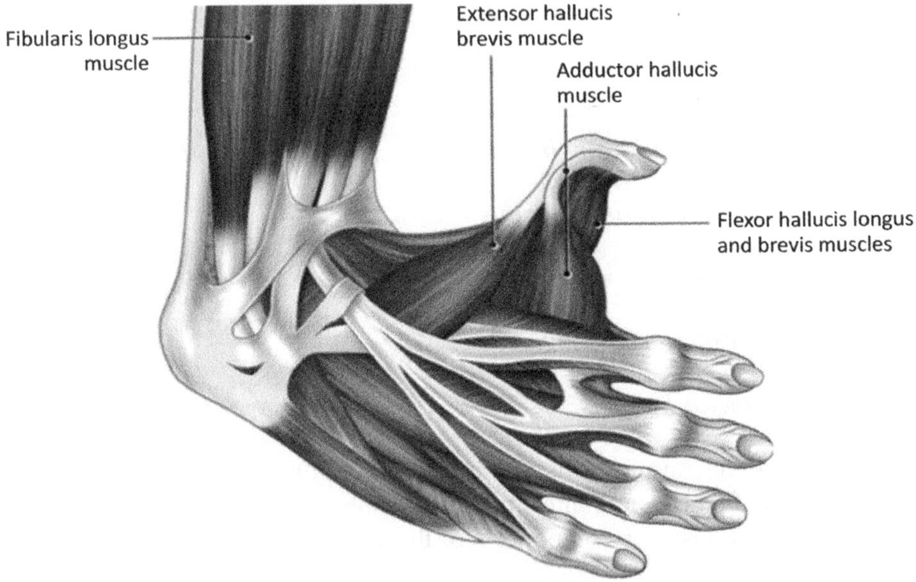

(b) Development of foot muscles (ape)

Figure 2.1B. *Short first ray muscles in humans relative to apes*

In comparison, these are overdeveloped in the great ape (Figures 1.20 and 2.1), as they are necessary for essential functions such as grasping and tree climbing. During a surgical procedure in humans, must these components be preserved at all costs to restore what is now a subsidiary function?

2.2. Hallux valgus

2.2.1. *Pathophysiology*

Hallux valgus, more commonly known as "bunion" (Figures 2.2 and 2.3), is a very frequent pathology which poses a semantic problem[1].

1 https://www.chirurgie-du-pied.fr.

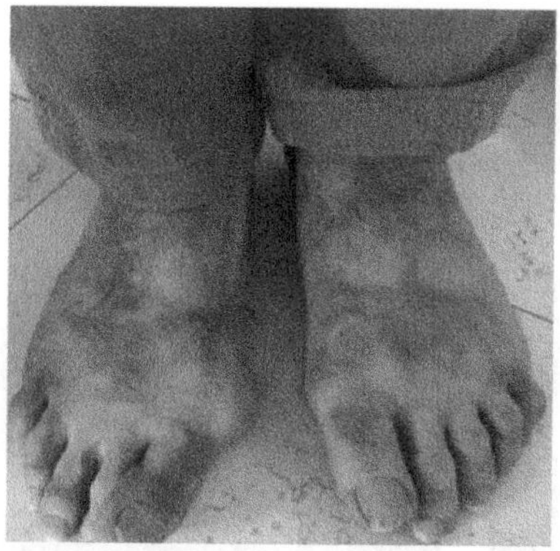

Figure 2.2. *Right hallux valgus with metatarsus varus*

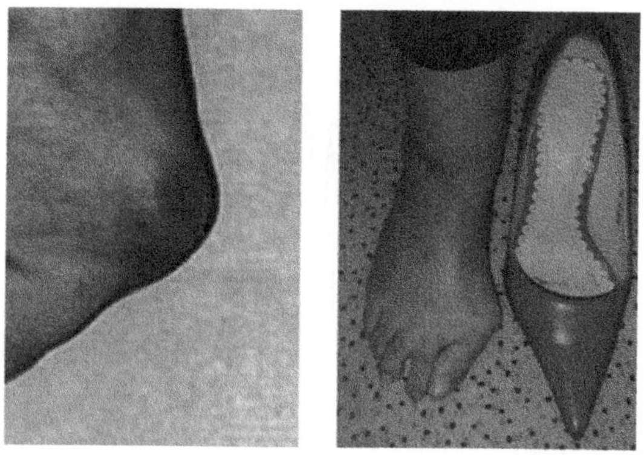

Figure 2.3. *Impingement in footwear*

Literally, having the hallux (big toe) in valgus (deviated outwards towards the little toe) is not a pathology; it is normal morphology and, above all, a specificity of the genus *Homo.* The pathology is never defined by the morphology, more so as it is normal, but by the symptoms that may be related to it. Humans are the only species to present a hallux in valgus, with a slight angular divergence (<9°) between the first and second metatarsals. The increase in this angle defines *metatarsus varus* (Figures 2.4 and 2.5), even if small in humans, and allows the maximum load to be placed under the first ray during the gait cycle, characteristic of human bipedalism.

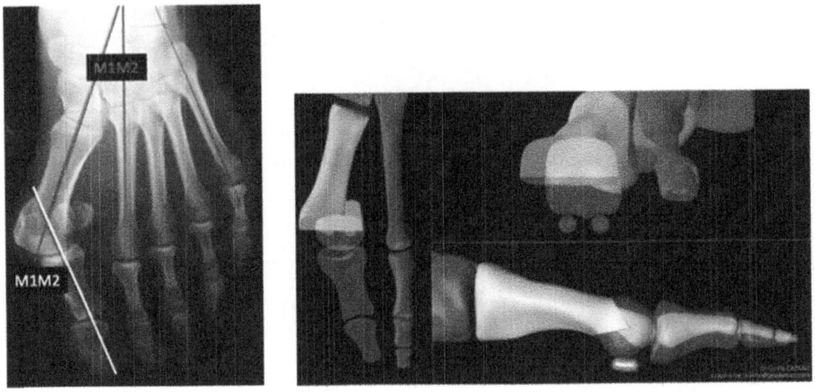

Figure 2.4. *Metatarsus varus (M1/M2) angle in human pathology. For a color version of the figure, see www.iste.co.uk/cazeau/foot.zip*

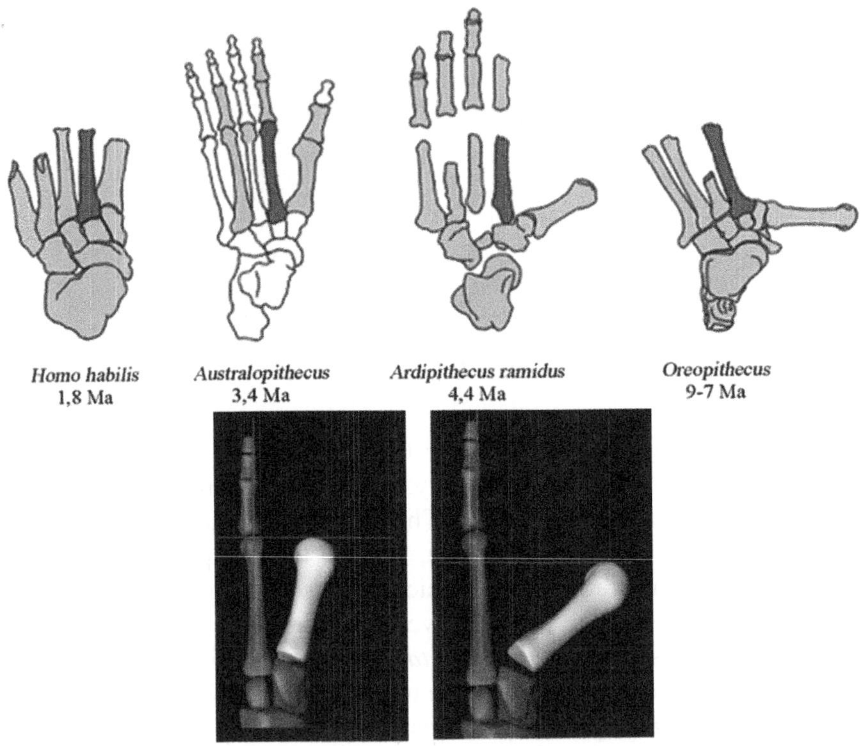

Figure 2.5. *Metatarsus varus of some fossil hominids, with 3D modeling. For a color version of the figure, see www.iste.co.uk/cazeau/foot.zip*

When this angle is increased (fossil hominids or pathological angle in *Homo* species) (Figure 2.5), the first ray can no longer bear the load and the body's weight is transferred to the lateral rays, a source of symptoms.

The greater the angle, the shorter the effective length of M1; this is called geometric brevity (Table 2.1). For example, an M1 with an actual anatomical length of 70 mm will have an effective, geometric length of 65.8 mm due to the cosine-related reduction coefficient of the M1/M2 angle of 20°.

	ANGLE M1/M2	10°	15°	20°	25°
Anatomical length	65 mm	64	62.7	61.1	58.9
	70 mm	68.9	67.6	65.8	63.4
		Geometric length			

Table 2.1. *Reduction in the useful length of the M1 (geometrical length when the M1/M2 angle increases is less than the anatomical length)*

Hallux deviation accompanies that of the first metatarsal (M1) constituting the hallux valgus (Figures 2.2–2.4); the big toe is no longer straight and cannot perform its propulsion functions needed for efficient human bipedalism. This is measured by the metatarsal-phalangeal angle (MTP1) (Figure 2.4).

Under these conditions where the first ray's stance function and hallux propulsion are compromised, we move away from human bipedalism to approach that of other primates.

2.2.2. *Normality/utility balance*

2.2.2.1. *Convergence towards normality and utility*

Surgery will focus on restoring function through realignment osteotomies. Reducing the angle between the first and second metatarsals (angle M1/M2), also called the *metatarsus varus* (Figures 2.4 and 2.5), restores an effective M1 length to eliminate the geometric brevity linked to the cosine of the angle M1/M2. In practice, a wide variety of techniques can be used, with the wide agreement about cutting (osteotomy) M1, moving its head laterally towards the little toe "in the foot", so as to put it back on its axis (Figure 2.6a and b).

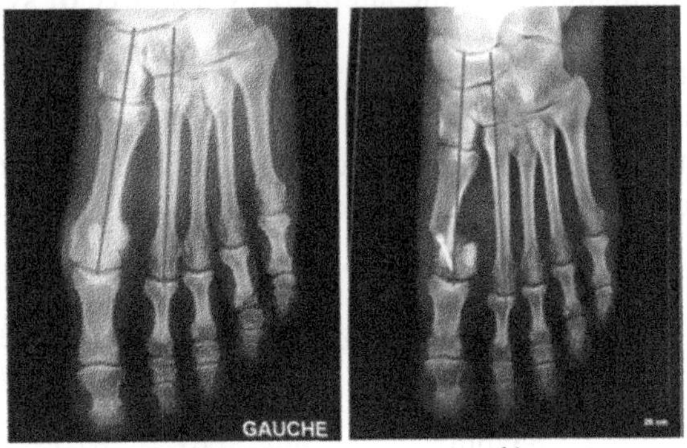

a) b)

Figure 2.6. *Distal osteotomy of M1 allowing lateral translation to reduce metatarsus varus. a) Preoperative radiograph. b) Postoperative radiograph. For a color version of the figure, see www.iste.co.uk/cazeau/foot.zip*

In this case, we can conclude that the surgical procedure has re-established normality and utility at the same time. The hallux is put back in the axis, restoring the propulsion function, decreasing the valgus moment and again approaching the situation of anatomical normality and functional utility (Figures 2.7 and 2.8).

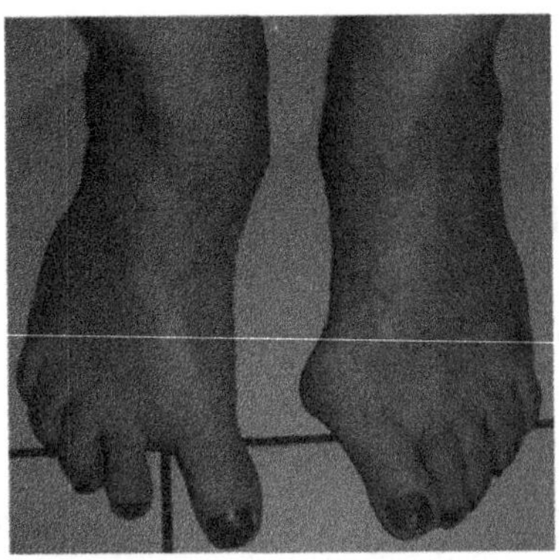

Figure 2.7. *Hallux varus on the right, valgus on the left; both alter the propulsion capabilities of the hallux*

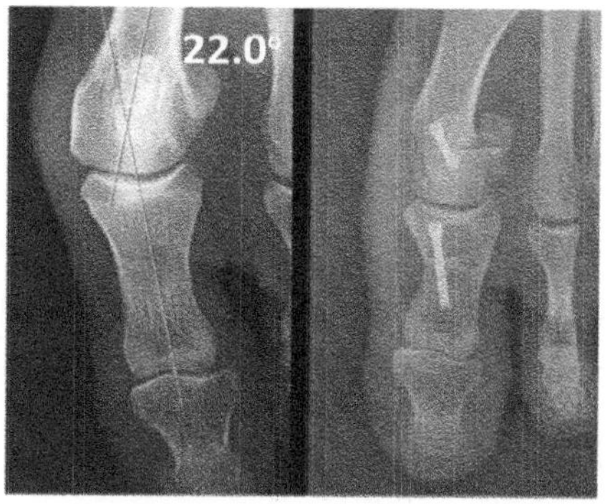

Figure 2.8. *Hallux realignment osteotomy restoring propulsion*

2.2.2.2. *Deviation from normality*

Certain surgical procedures can deviate from normality such as arthrolysis of the first metatarsophalangeal joint. To help redirect the hallux, the retracted joint elements on the lateral side should be relaxed (Figure 2.9).

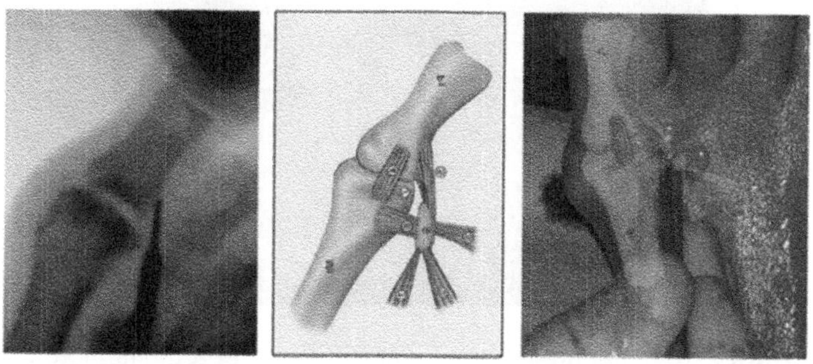

Figure 2.9. *Lateral arthrolysis of MTP1, by tenotomy of the retracted elements. For a color version of the figure, see www.iste.co.uk/cazeau/foot.zip*

This requirement cannot be ignored, no matter which surgical technique is used. The anatomical elements present define our normality. However, certain components must be cut to restore the function of utility. The functional importance of the anatomical elements present is listed and classified hierarchically, and thanks to the prism of comparative anatomy, it is quickly understood that certain components (adductor, transverse and oblique components) are present but useless in *Homo* species, given the absence of grasping, first ray opposition, brachiation and quadrupedalism (Figure 2.1).

The opposite is true for many apes, in whom this anatomical sacrifice would be impossible.

2.3. Arthrodesis

Arthrodesis is the surgical fusion of a joint because of a severe joint destruction (Figure 2.10) or joint instability; it results in loss of function. The examples illustrated here involve the first metatarsophalangeal, cuneometatarsal and interphalangeal joints of the toes[2].

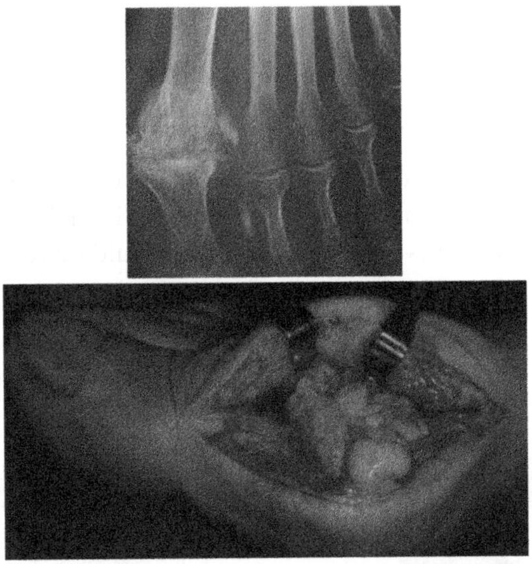

Figure 2.10. *Joint destruction of MTP1 by osteoarthritis; note the exuberant bone production (osteophytes). For a color version of the figure, see www.iste.co.uk/cazeau/foot.zip*

2.3.1. *The first cuneometatarsal joint (C1M1)*

2.3.1.1. *Introduction*

In humans, the midtarsal area is interlocked with the base of the second metatarsal, which is fundamentally opposed to the midtarsal break in apes (Figure 1.30). The base of the first human first ray is neither mobile nor opposable. Dorsiflexion does not take place in the midtarsal area but more

2 https://www.chirurgie-du-pied.fr.

distally in the first metatarsophalangeal joint (MTP1), which is very mobile, especially in the extension (Figure 1.31). Under these conditions, C1M1 pathological joint instability is contrary to human bipedalism, and its fixation, known as arthrodesis, is close to the criteria of anatomical normality and functional utility[3].

2.3.1.2. *Anatomical review*

The anatomical means of union between the first cuneiform and the first metatarsal is primarily the dorsal ligament. It is the thickest structure on the dorsal-medial side. The plantar ligament has a larger surface area, is rectangular in shape and has a rather lateral topography. Plantar release of the joint on its lateral side is important to mobilize M1. There is no purely medial or lateral ligament.

The anterior surface of the first cuneiform has a reniform shape that is convex medially and concave laterally. It is sometimes restricted, even constituting two articular surfaces. The orientation is oblique anteriorly downwards and inwards. The shape is discreetly convex transversely, flat vertically.

The posterior surface of the first metatarsal has a triangular lower base, its major axis is oriented downward and outward, its shape is flat vertically. Two insertions are noteworthy: the anterior tibial tendon on its inferior-medial side and the long fibular tendon on its inferior-lateral side. Its rearward orientation varies from 13° medially to 13° laterally.

On the lateral aspect of the M1 base, the articular surfaces connected to the second metatarsal are not consistent. These surfaces are a valuable, albeit indirect, paleontological indicator of the degree of M1 opposition.

The advantage of attaching the first ray to the second ray, for example, through the first metatarsal to the second metatarsal, is linked to the second metatarsal's stability. Indeed, locking the base of the second metatarsal in the tarsus, articulated with M1 M3, C1, C2 and C3, allows locking of the midtarsal area, and is specific to the genus *Homo*.

2.3.1.3. *Clinical presentation*

Load transfer to the lateral rays can be a direct consequence of C1M1 laxity. It must be systematically detected; missing it leads to surgical failure.

---

3 https://www.chirurgie-du-pied.fr.

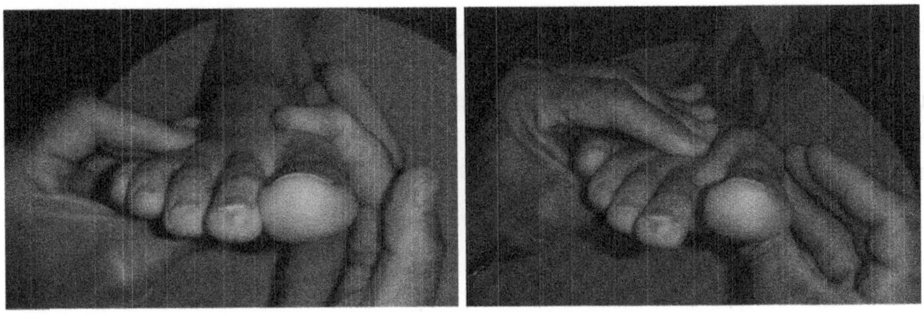

Figure 2.11. *C1M1 instability: clinical drawer test showing vertical joint instability*

Clinically, one hand is used to grasp all the lateral rays, while the other hand tests C1M1 mobility in dorsal and plantar translation instead of flexion and extension (Figure 2.11). The other useful sign is the loss of MTP1 extension when the first metatarsal joint is elevated (hyperextension of C1M1).

Signs of foot instability can also be observed, especially in single-leg stance, as the patient feels he "can no longer rely on his foot". This finding of instability is even more valuable when the other causes are absent (valgus hindfoot, cerebral-vestibular and cerebellar senescence). This laxity is sometimes generalized, physiological or due to certain diseases (Figure 2.12).

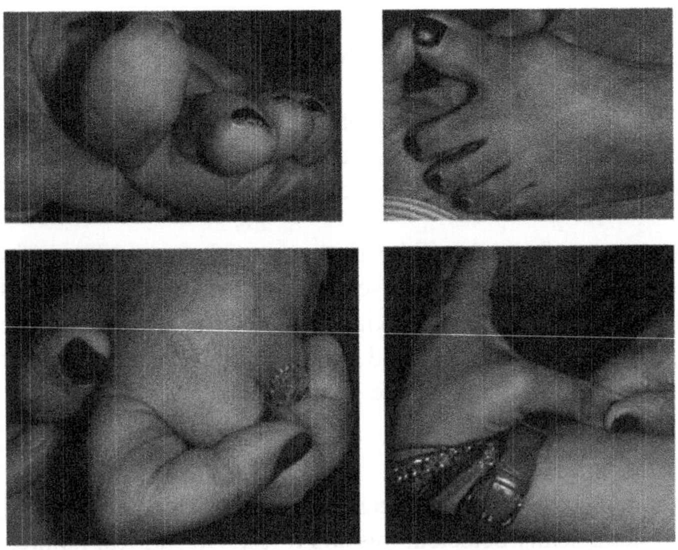

Figure 2.12. *Widespread instability in the Ehlers Danlos disease*

This results in a loss of anatomical ratios on CT scans when the foot is loaded (Figure 2.13).

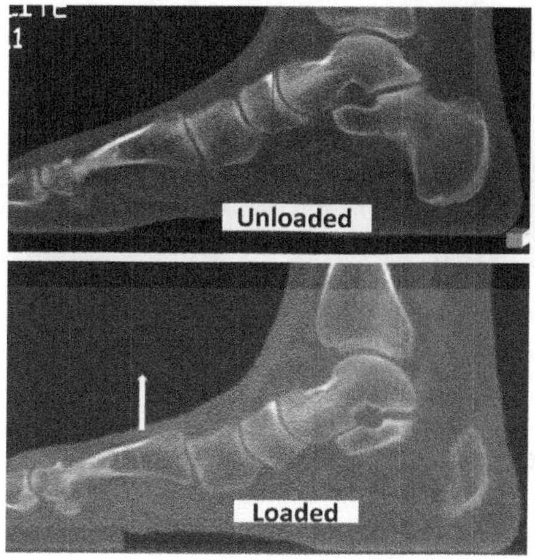

Figure 2.13. *CT scan with load bearing reveals C1M1 instability (loss of normal joint ratios), plantar joint gapping and dorsal translation of M1*

2.3.1.4. *Surgical treatment*

The surgical intervention locks the C1M1 joint, eliminating pathological mobility and load transfer to the lateral rays. Thus, it resembles the normal model of human bipedalism. Figure 2.14 shows the minimally invasive technique, replacing the overly aggressive classical technique (Figure 2.15).

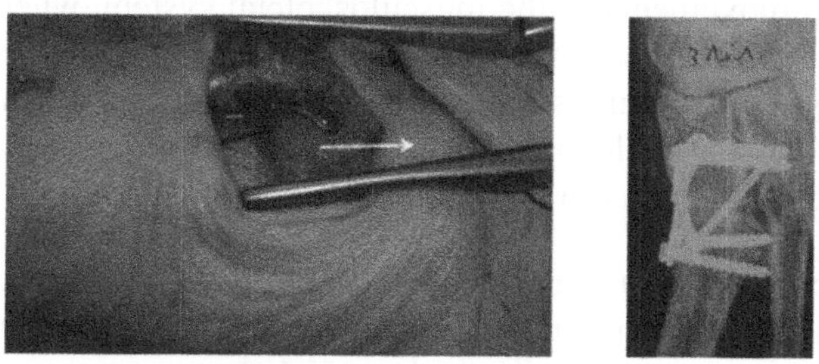

Figure 2.14. *C1M1 arthrodesis; placement of a titanium plate to fuse the joint and eliminate instability. For a color version of the figure, see www.iste.co.uk/cazeau/foot.zip*

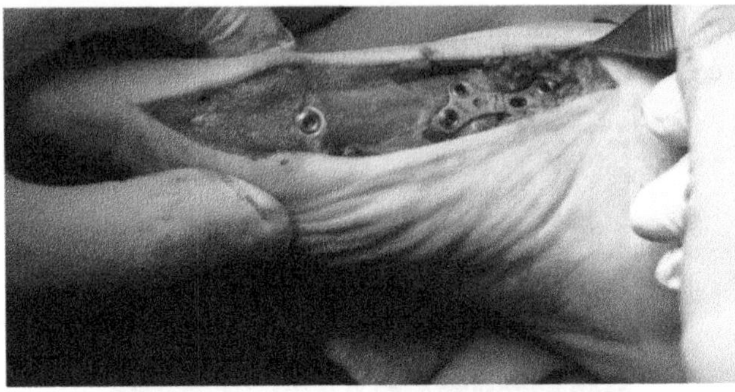

Figure 2.15. *C1M1 arthrodesis via a conventional, extensive open approach. For a color version of the figure, see www.iste.co.uk/cazeau/foot.zip*

2.3.2. *First metatarsophalangeal joint (MTP1)*

2.3.2.1. *Biomechanics, the articular chain*

MTP1 has a 30° mobility in plantar flexion and 90° in dorsal flexion (Figure 1.30). It is mainly in extension in the human species. It is the flex point of the foot in humans, in association with the other links of the articular chain constituted by the talonavicular, cuneometatarsal and interphalangeal joints of the hallux. This is in contrast to non-human primates, whose flex point is located at the midtarsal break (Figure 1.31a and b).

In humans, there is hyperextension of the MTP joints and locking of the midtarsal joint by interlocking with the base of the second metatarsal.

As with all movements of the musculoskeletal system, whether the upper limb, lower limb or spine, an entire articular chain is brought into play and not just a single joint. This is not copiously seen macroscopically, except in animals with an exoskeleton. This is evident from the movement of an arachnid appendage or a crab leg (Figure 2.16a and b).

Under these conditions, it is understandable that, within an articulated chain, neutralizing one joint will not block the whole chain, but instead transfer the stresses to the other articulated links[4].

4 https://www.chirurgie-du-pied.fr.

Figure 2.16. *Multi-articulated segment of an arthropod appendage making up an articular chain. a) Theraphosa blondi. b) Crustacean*

2.3.2.2. *Pathophysiology*

Those suffering from MTP1 osteoarthritis may benefit from early conservative surgical procedures to reduce pain and maintain residual mobility.

For example, the M1 and hallux can be shortened to reduce the intra-articular MTP1 pressure (Figure 2.17a and b).

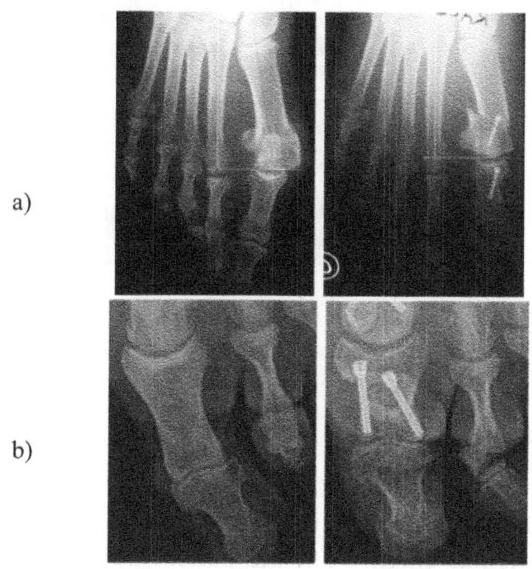

Figure 2.17. *a) Surgical shortening of M1. b) Surgical shortening of hallux*

The gait cycle is also aided by performing staged osteotomies that place the bones in dorsal flexion (Figure 2.18).

When the disease is very advanced, it is treated by blocking the joint (arthrodesis) (Figure 2.19). This has the effect of completely eliminating pain and bone production of osteophytes.

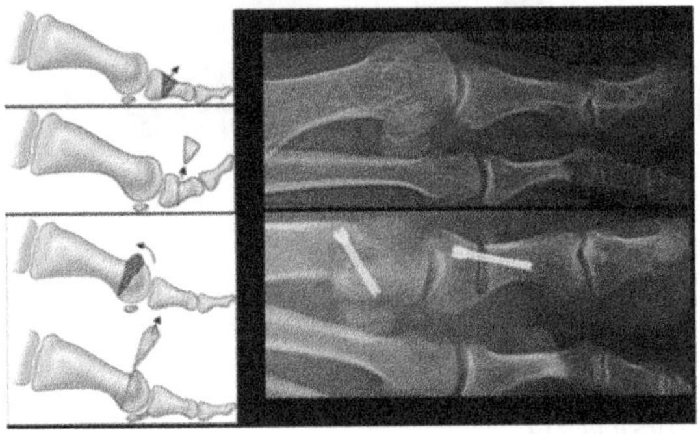

Figure 2.18. *Osteotomies of M1 and P1 in dorsal flexion, to reset the flex point. For a color version of the figure, see www.iste.co.uk/cazeau/foot.zip*

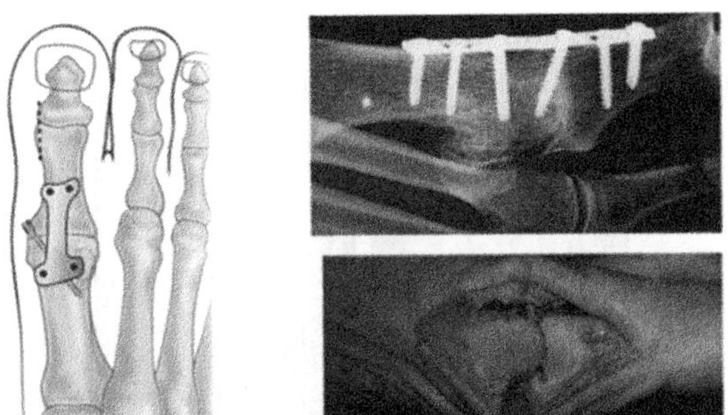

Figure 2.19. *Arthrodesis of MTP1 using a titanium metal plate. For a color version of the figure, see www.iste.co.uk/cazeau/foot.zip*

2.3.2.3. *Surgical treatment*

How can function be restored in advanced osteoarthritis? MTP1 pain and stiffness prevent its use at the flex point, as well as reduce the use of the other components of the articular chain, with the patient adopting an altered gait pattern to avoid pain. This is often done in varus, the foot resting on its lateral edge as in apes.

It is not possible to return to a state of anatomic normality. The arthritic MTP1 cartilages are too damaged; they are neither repairable nor replaceable by a reliable prosthesis. What can be done to help recover gait function? It is useful to restore the function of the other three joints of the articular chain by sacrificing the weak, painful link.

This is the principle of arthrodesis (Figure 2.19). By removing painful residual mobility, the rest of the chain is allowed to function properly. This is a departure from the original normality model in favor of functional utility. Excessive dorsal flexion when adjusting the phalanx is detrimental, as it prevents the user from walking flat (Figure 2.20).

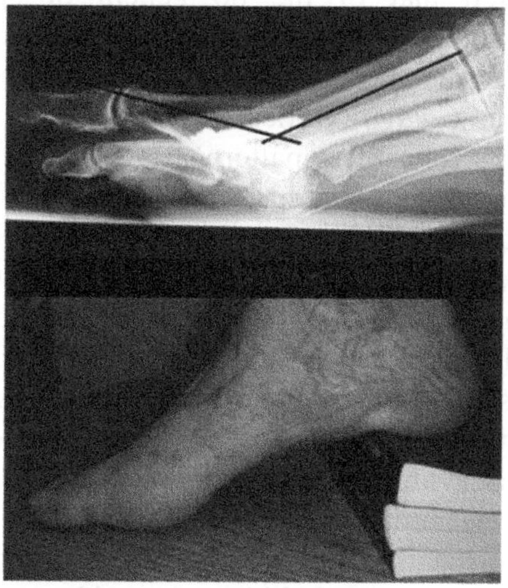

Figure 2.20. *MTP1 arthrodesis with excessive dorsal flexion adjustment; to be flat on the ground, the heel must be very elevated*

2.4. Excessive mechanical pressure under the forefoot (second, third and fourth metatarsals – M2, M3 and M4, respectively)

2.4.1. *Biomechanics and evolution*

In 1974, a comprehensive analysis of metatarsal robustness was done by Riesenfeld. What follows are his results and our observations from his data.

The main results of Riesenfeld's (1974) study do not take into account the second metatarsal and reflect the lesser importance of M2 with respect to M1 and M5 in classifying the robustness of metatarsals in primates, rodents and insectivores.

The mean signature 1-5-4-3-2, which designates the order of the five metatarsals (M1, M5, M4, M3, M2) from the most robust to the least robust, is unique in humans. It has previously been shown that this order of robustness varies in humans, *Pan* and *Gorilla* (Archibald *et al.* 1972).

Among the other great apes, four have M5 as the most slender of the metatarsal group (*Hylobates*, *Symphalangus*, *Pan*, *Gorilla*), while *Pongo*'s M5 is the third strongest metatarsal. Of the 51 genera of primates described by Riesenfeld (1974), 33 genera (64.7%) have M5 as the most slender. It should also be noted that of the 86 genera or species of primates, insectivores and rodents he studied, only *Cynomys* rodents do not have the M1 as the most robust metatarsus, since M5 holds this position. For the same sample, only the genus *Galago* (vertical suspensor and jumper) is similar to the genus *Homo* in its robustness pattern. Riesenfeld (1974) focused mainly on M1 and M5. He concluded that his data link the 1 + 5 robustness model (M1 being the most robust and M5 the second most robust) with terrestrial displacement and the 1 → 5 model (M1 being the most robust and M5 the least robust of the metatarsals) with arboreal displacement. Robustness can be more an indicator of substrate than of a particular type of locomotion. Thus, the human robustness signature refers primarily to the substrate over which we move. The existence of the 1 + 5 robustness model in various groups (including humans) is evidence that primitive traits are retained in present-day humans. However, human intraspecies variation indicates a multitude of relative rankings of metatarsal robustness that is not compatible according to Riesenfeld (1974) with what it would have been if our robustness model had been inherited from terrestrial mammals. It should be noted that *only* 53.4% have the 1 + 5 model in humans. The arboreal model (1 → 5) is present in 16.9%.

According to Riesenfeld's conclusions (1974), the 1 + 5 model of human robustness is the consequence of adaptation to the terrestrial substratum; therefore, it can be incorporated into the analysis of fossil parts and to help us understand the evolution of human locomotion.

2.4.2. *Pathophysiology*

Excessive pressure under the M2, M3 heads, and to a much lesser extent M4, quickly results in plantar pain, due to the fixed nature of the tarsal–metatarsal joints at this level. This excessive pressure has many causes, both in the foot and the body.

With respect to the foot, functional impairment of the first ray (Figures 2.4 and 2.5) is often due to *metatarsus varus*. It can also be a pes cavus (hollow) foot (Figures 2.21–2.23) due to verticalization of the metatarsals, equinus deformity due to retraction of the gastrocnemius and many other causes.

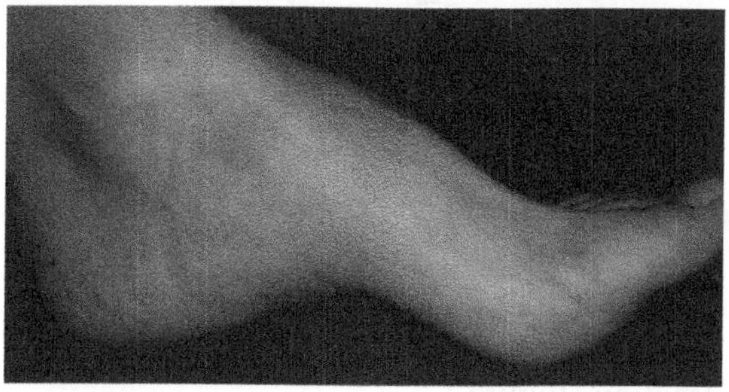

Figure 2.21. *Hollow foot, clinical appearance*

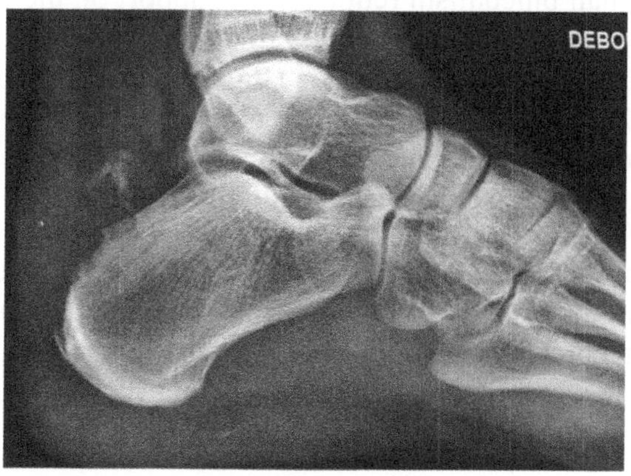

Figure 2.22. *Hollow foot, radiological appearance*

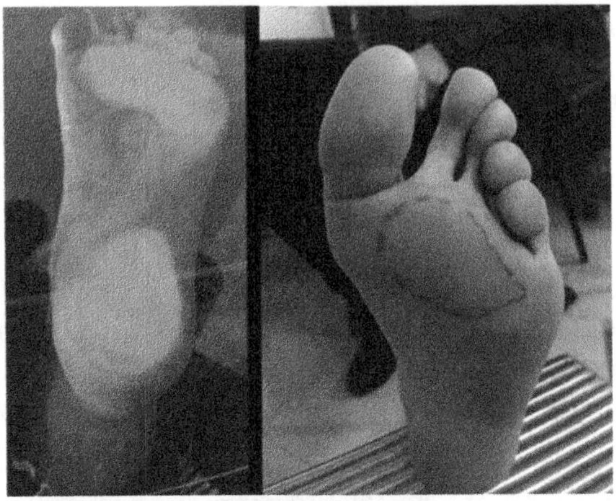

Figure 2.23. *Hollow foot; plantar impression, hyperkeratotic zone due to excessive pressure. For a color version of the figure, see www.iste.co.uk/cazeau/foot.zip*

2.4.3. *Surgical treatment*

Treatment is aimed at the cause. The treatment of *metatarsus varus* in hallux valgus has been prescribed (Figure 2.6a and b), including gastrocnemius retraction (Figure 1.27).

Excessive pressure is treated for itself in disabled patients by removing the function; human bipedalism requires full support on the metatarsal heads at the end of the stance phase. To make this phase painless, osteotomies are performed to raise the metatarsal heads (Figure 2.24), which, although they deviate from the model of anatomical normality, are functionally useful because they reduce plantar[5] pressures. This helps to recreate the transverse arch of the foot in a spectacular manner (Figure 2.25).

The harmony of the metatarsal cascade should be preserved, with a decrease in length from M2 to M5, even if this is only a geometric approximation taking into account the true length and the apparent length depending on the obliquity of the bone relative to the incident rays (Figure 2.26).

5 https://www.chirurgie-du-pied.fr.

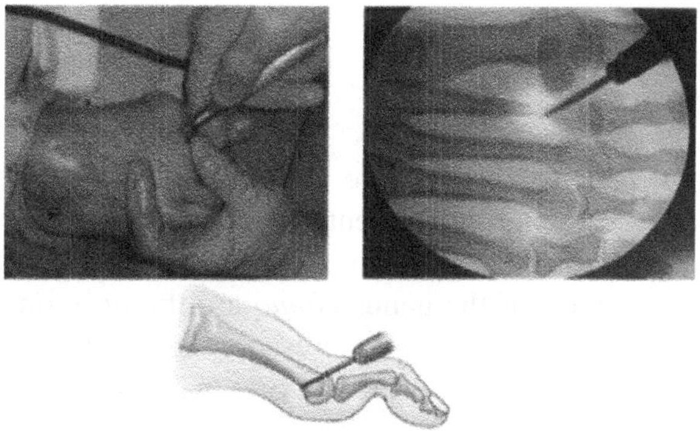

Figure 2.24. *Percutaneous osteotomies of M2, M3 and M4 for metatarsalgia due to excessive pressure on the forefoot (distal metatarsal mini invasive osteotomy, DMMO)*

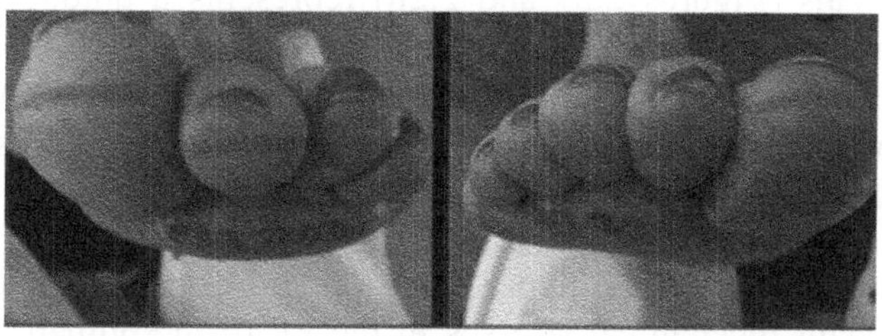

Figure 2.25. *Left, the collapsed plantar arch is convex downward (round forefoot); restored on the right by percutaneous metatarsal head-raising osteotomies (DMMO)*

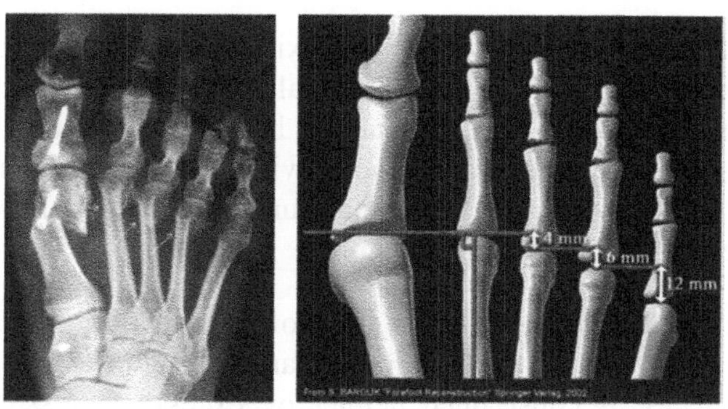

Figure 2.26. *Radiograph showing the surgical correction of the first ray and osteotomies to raise the lateral rays; the metatarsal cascade is preserved according to the diagram on the right. For a color version of the figure, see www.iste.co.uk/cazeau/foot.zip*

2.5. Plantar plate

2.5.1. *Introduction*

The plantar plate (PP) is a passive structure that is not heavily studied. It is part of the capsuloligamentous complex that stabilizes the metatarsophalangeal joints. It sits in a complex anatomical and dynamic environment. The species of the genus *Homo* are the only ones that undergo double mechanical stress, excessive pressure and hyperextension, which cause lesions whose repair requires making a choice between return to normality and return to function.

2.5.2. *Plantar plate in primates*

PP in humans (Figures 2.27 and 2.28) represents a unique situation not found in any other living being. The forefoot of the genus *Homo* is characterized by highly mobile metatarsophalangeal joints (MTPs), especially in extension, and locked tarsal–metatarsal joints, especially the second and third rays (Figure 1.31).

This combination of MTP hyperextension and excessive plantar pressure places significant mechanical stresses in the PP of the human foot. Even in terms of the human palmar plates, the mechanical conditions are different because of the absence of metacarpophalangeal hyperextension and the lack of support.

In apes (Figure 1.32a), MTP dorsal flexion is limited due to the plantar orientation of the cartilages of the metatarsal heads, and by the presence of a large protrusion behind the metatarsal head. In addition, the lateral metatarsals have an inversion torsion, with the heads facing the first metatarsal, emphasizing the grasping function, which is opposite to permanent terrestrial human bipedalism.

Conversely, the tarsal–metatarsal joints of apes are very mobile, in dorsal flexion, flex point and the ankle is in permanent dorsal flexion. Under these conditions, the PP is constrained neither by excessive pressure under the metatarsal heads nor by stretching since there is no MTP hyperextension.

2.5.3. *Anatomy*

Four anatomical structures are critical to PP biomechanics:

– the PP itself, the plantar fascia, the collateral ligaments of the MTP joints and the flexor digitorum longus (FDL) tendon;

– the PP joins the head of M1 and the base of P1 on the plantar side of the foot (Figure 2.27). Here, a sagittal anatomical section shows head extension with the cartilaginous surface of the base of the first phalanx.

a)

b)

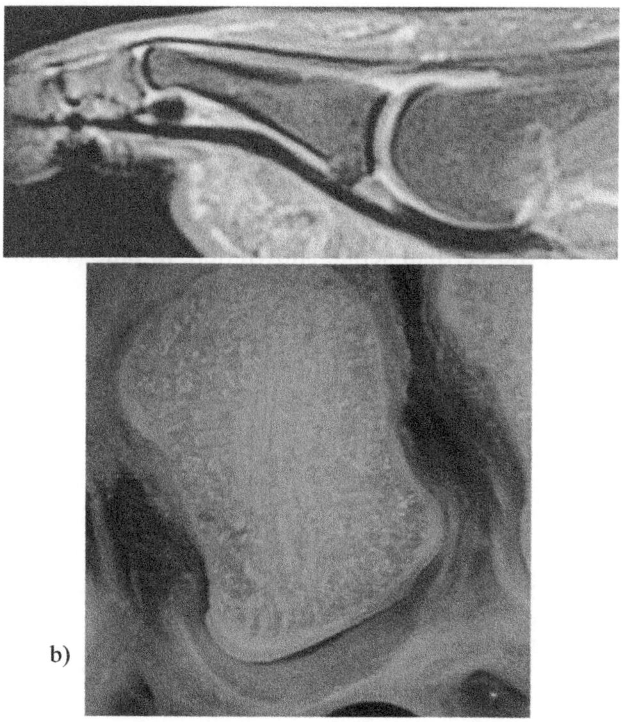

Figure 2.27. *Plantar plate viewed in different modalities. a) MRI sagittal section and b) frontal cadaveric section*

Its fibrocartilaginous nature and the strong phalangeal insertion are noteworthy (Figure 2.28).

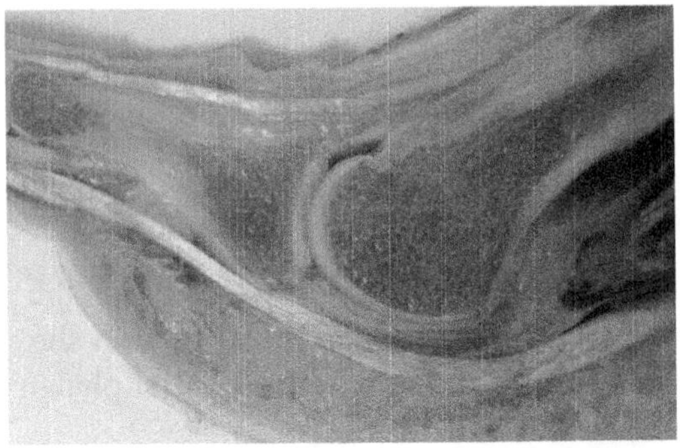

Figure 2.28. *Cadaveric sagittal section showing articular extension of the base of the hallux by the plantar plate (source: Prof. Xavier Demondion – we thank Prof. X. Demondion of the Laboratory of Anatomy at the University Hospital of Lille for these very high quality preparations)*

The PP and the superficial plantar fascia are anatomically linked through the intermediary of the vertical septa, shown here in blue (Figure 2.29).

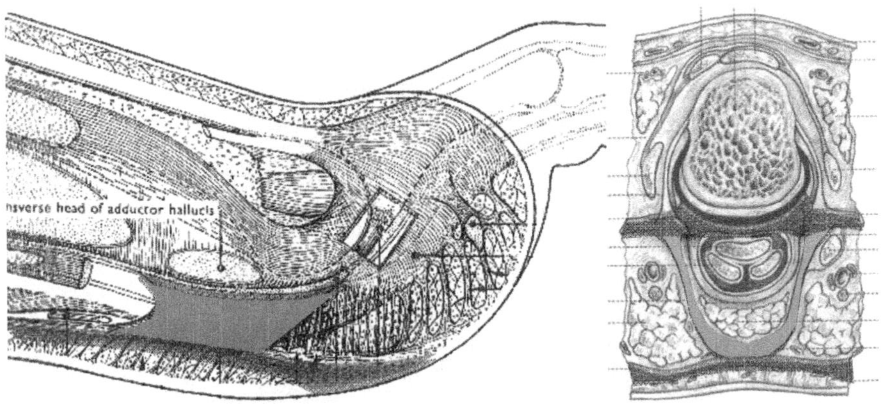

Figure 2.29. *Plantar fascia and its connections with the plantar plate. For a color version of the figure, see www.iste.co.uk/cazeau/foot.zip*

As for the collateral ligaments (Figure 2.30), unlike the first ray, the metatarsoglenoid ligament is of little importance, mechanically weak, and the collateral ligaments are more robust, the fibular one in particular.

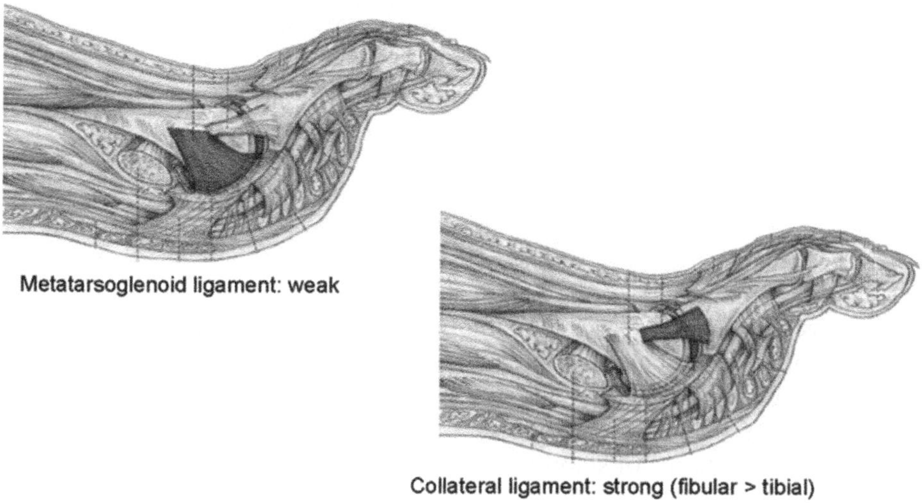

Figure 2.30. *Metatarsophalangeal collateral ligaments. For a color version of the figure, see www.iste.co.uk/cazeau/foot.zip*

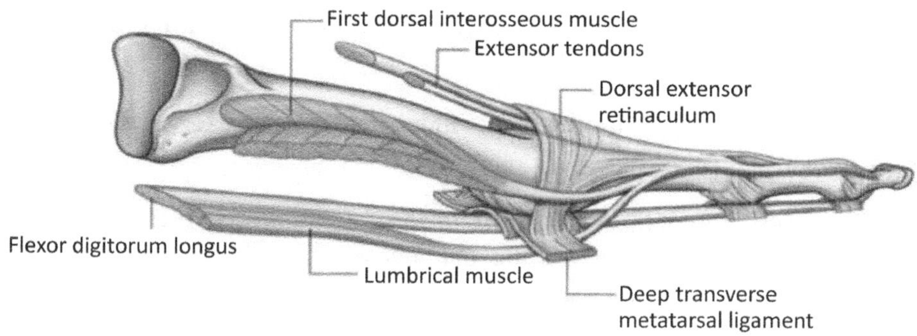

Figure 2.31. *Flexor digitorum longus. For a color version of the figure, see www.iste.co.uk/cazeau/foot.zip*

Finally, as the last anatomical player, the FDL (Figure 2.31) is longer and more powerful than the flexor digitorum brevis (FDB). It has several interesting characteristics. It is a plantar flexor of the MTPs, PIPs and DIPs (proximal and distal interphalangeal joints), due to its plantar insertion on the distal phalanx. Located some distance from the point of rotation of the MTP, it produces larger moments. The FDL is attached to the lumbrical muscles (*musculi lumbricales pedis*), which are themselves plantar flexors of the MTP.

Thus, the FDL is the main plantar flexor of the MTPs.

Among the anatomical structures cited, only the FDL is active, the others are passive.

2.5.4. *Biomechanics*

2.5.4.1. *Passive structures*

Cadaver studies of selective tenotomy have improved our understanding of the relative importance of these passive anatomical structures. Only the selective cutting of the PP results in a significant difference in stability. This is not the case for isolated tenotomy of the collateral ligaments, which are less important in terms of stability.

On the other hand – and this is a very interesting result – the combined transection of the PP and one or both collateral ligament(s) causes the greatest loss of joint stability.

We will attempt to chronologically determine the critical moment when the PP is under the most significant stress.

Answering this question helps us understand what preserves the PP and what damages it.

The critical moment corresponds to the maximum MTP extension, i.e. chronologically to the part of the gait cycle when the pressure is at its maximum under the metatarsal heads.

This is determined by the dynamic study curves representing the center of gravity (CG) trajectory.

At the flex point, the CG is lowered and then the peak propulsion is observed. The centers of pressure are then located under the metatarsal heads; it is at this point that the MTP joints will go into maximum dorsal flexion. This zone, called zone C, is located between the 40th and 60th percentiles of the gait cycle (Figures 1.21 and 1.22).

2.5.4.2. *Active structures*

What about the active structures, i.e. the FDL? It is very interesting to note that it is also during zone C that the FDL and FHL (*flexor hallux longus*) activation occurs (Figure 2.32). Thus, it probably has an active braking and stabilizing role in the MTP.

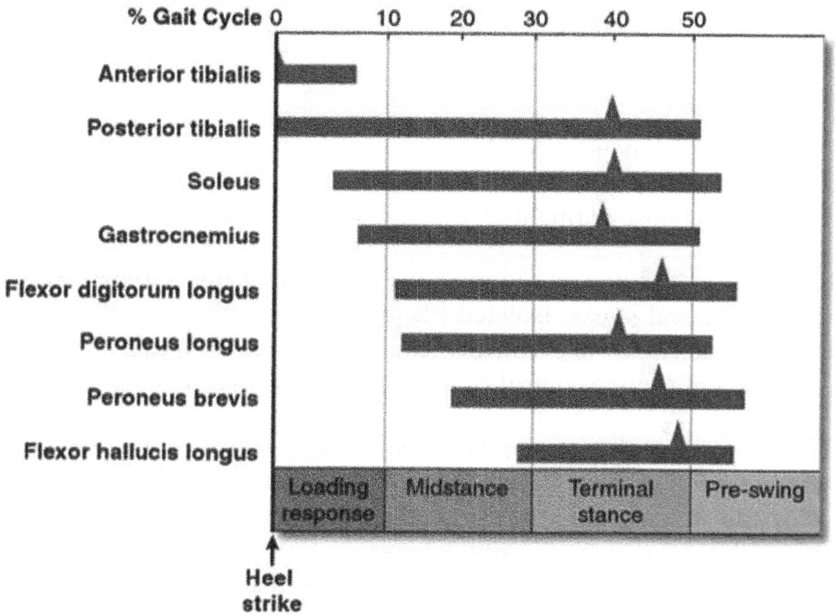

Figure 2.32. *FDL and FHL activation between the 40th and 60th percentiles of the gait cycle (source: Perry (1992))*

2.5.4.3. *In total*

There is simultaneous MTP hyperextension, maximum pressure under the metatarsal heads, maximum stretching of the PP and FDL activation at the critical moment.

The three passive structures thus share the strain. The FDL is the only active structure capable of protecting the PP. This is a unique phenomenon in humans due to the specificity of the distal flex point, which is absent even in the human hand.

2.5.5. *Mechanical balance*

Among the structures stabilizing the MTP joint, the PP has an essential role. Its destabilizing action is even more important when the collateral ligaments are damaged. The FDL tendon has a protective role.

The PP is most at risk mechanically with the combination of excessive pressure under the metatarsal heads, and the tensile stresses associated with the flex point.

2.5.6. *Treatment*

Should we choose normality?

It would be a matter of suturing the structures damaged because of the double mechanical stress. Isolated PP repair returns the system to its initial unstable situation and will eventually lead to a recurrence. Weakening the FDL removes the only active control element of the system.

The choice is utility to the detriment of anatomical normality: the only geometric solution that keeps the MTP's dorsal flexion amplitude by reducing the weight-related loads on the PP is shortening and elevating the metatarsal heads, which is easily achievable by DMMO (Distal Metatarsal Mini-invasive Osteotomy). The damaged structures are not restored, but the cause – excessive pressure – is removed and the system is relaxed by shortening, thus restoring the function without preserving the original anatomy.

2.6. Release of gastrocnemius muscles

Contracture of the gastrocnemius muscles during walking results in a loss of ankle dorsiflexion, especially when the knee is extended (Figures 1.21 and 2.33).

This generates excessive pressure under the metatarsal heads when the swing phase (toe off) occurs (Figure 1.22). The surgical choice is as follows.

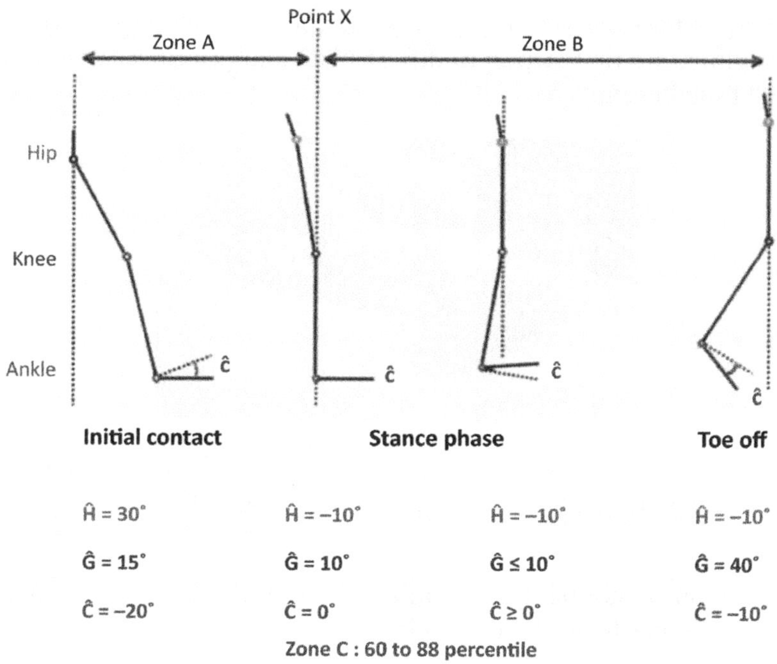

Figure 2.33. *Kinematics: linkage modeling of the stance phase in humans. For a color version of the figure, see www.iste.co.uk/cazeau/foot.zip*

Normal anatomy can be restored by lengthening the gastrocnemius muscles (Figure 1.27), or gait function can be restored by lateral ray-raising osteotomies to remove the pain-causing excessive pressure (Figure 2.24).

Thus, in this example, healing can be achieved by choosing to either return to anatomical normality or preserve function (utility).

2.7. Claw toes (Figure 2.34)

2.7.1. *Surgical alternatives*

When faced with clawed toes, like for the entire locomotor system, the surgical options are to act on the soft tissues, the bones or the joints[6].

6 https://www.chirurgie-du-pied.fr.

Nothing is possible without making sacrifices. To be most effective and least crippling, the first step is to differentiate and prioritize what separates “normal“ from “useful“.

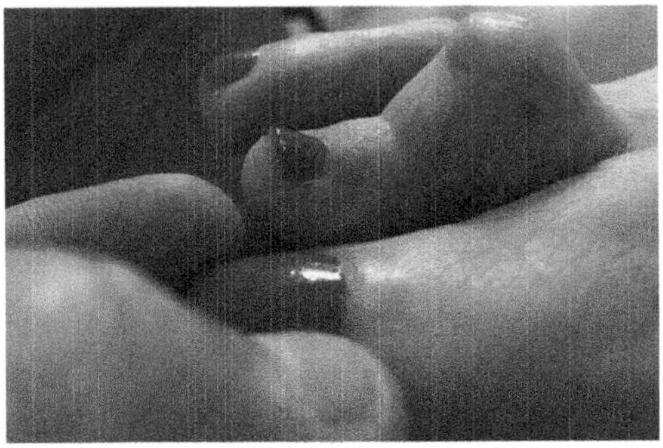

Figure 2.34. *Claw toe; MTP hyperextension, PIP flexion, fixed DIP recurvatum*

This notion of normality and utility (Figure 2.35) is very relative and depends on the mechanical model being used.

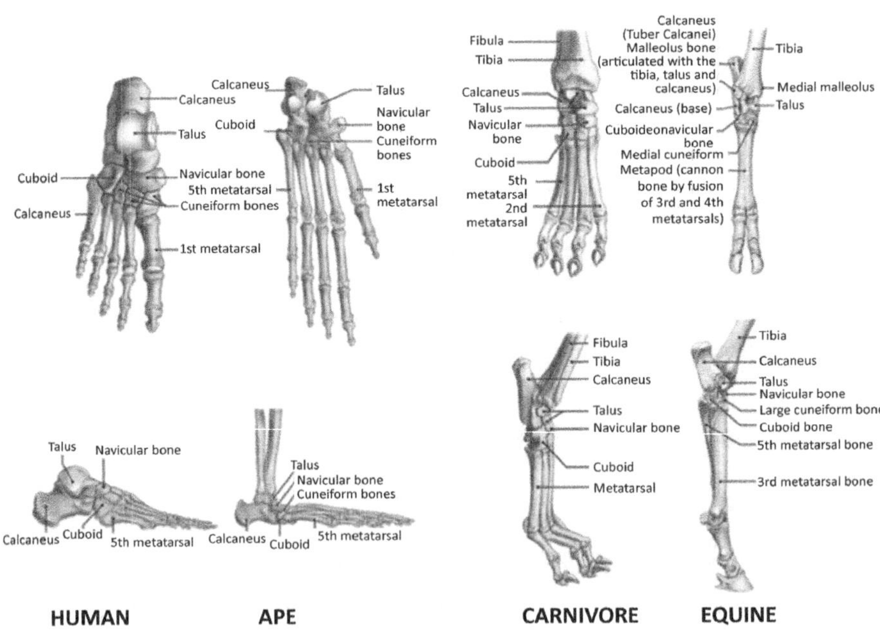

Figure 2.35. *The concept of normality differs depending on whether the animal has plantigrade (human, ape), digitigrade (carnivore) or unguligrade (horse) gait*

For example, in a plantigrade subject such as a human (*Homo* species), toe contact with the ground is not as significant as in an animal whose foot rests on the ground only through the toes, defining the specific digitigrade of carnivores such as dogs or cats. This is even different in ungulates such as horses, where ground contact occurs only by the nails, represented by hooves.

Evolution has produced specific adaptive modifications from one animal group to another; this allows us to distinguish a useful structure – indispensable to function – from a superfluous evolutionary residue that characterizes our "normal" anatomy.

2.7.2. *Anatomy reminder*

The various active motors and their functions are listed in Table 2.2 and Figure 2.36 (the main functions are shown in capital letters in Table 2.2). Anatomical details are given in Section 2.7.7.

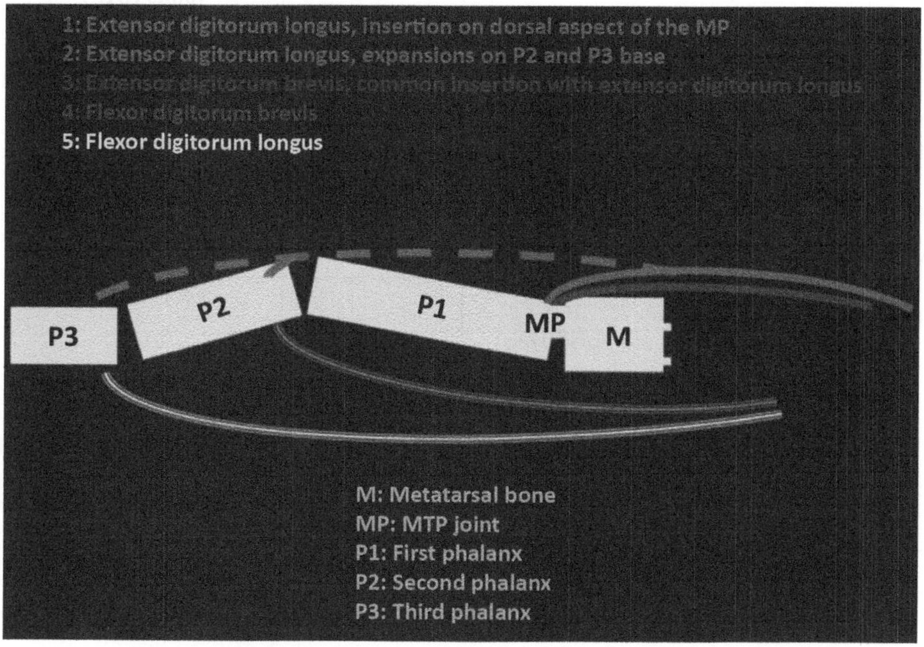

Figure 2.36. *Diagram of the insertions of the extrinsic foot muscles. For a color version of the figure, see www.iste.co.uk/cazeau/foot.zip*

NOTE.–

EDB = *extensor digitorum brevis*

EDL = *extensor digitorum longus*

FDB = *flexor digitorum brevis*

FDL = *flexor digitorum longus*

Dorsal interossei = *musculi interossei dorsales pedis*

Plantar interossei = *musculi interossei plantares*

Lumbrical muscles = *musculi lumbricales pedis*

	Metatarsophalangeal joint (MP)	**Proximal interphalangeal joints (PIP)**	**Distal interphalangeal joints (DIP)**
Dorsal interossei	FLEXION	Extension	Extension
Plantar interossei	FLEXION	Extension	Extension
Lumbrical muscles	FLEXION	Extension	Extension
Extensor digitorum brevis (EDB)	EXTENSION	0	0
Extensor digitorum longus (EDL)	EXTENSION	0 +++	0 +++
Flexor digitorum brevis (FDB)	Flexion	FLEXION	0
Flexor digitorum longus (FDL)	Flexion	Flexion	FLEXION

Table 2.2. *Function of the various anatomical structures*

2.7.3. *Extension function*

2.7.3.1. *Comparative anatomy*

Should the extension function be preserved? Extension at the metatarsophalangeal joints (MTPs) is a human specificity. This is not the case for great apes, where the mobility is shifted backwards at the midtarsal break (Figure 1.32a and b). One's "normality" is therefore not the same as the other's, even if both are lumped into the fuzzy functional anatomy concept of bipedalism.

It is interesting to note that, in the genus *Homo*, entrapment of the base of the second metatarsal (clinically responsible for the second ray syndrome) locks the distal part of the mediotarsus and moves the zone of mobility distally (Figure 1.31). Conversely, the large amount of mobility in extension of the great ape's midtarsal area is accompanied by locking of MTP extension. This is related, on the one hand, to the plantar orientation of the cartilaginous surface of the metatarsal heads, and, on the other hand, to the presence of a dorsal bony stop at the heads that prevents the phalanxes from lifting (Figure 1.32a and b).

We can then conclude that MTP extension is specific and essential in humans: it must be preserved for the foot to flex at the proper point.

2.7.3.2. *Extensor mechanism*

Should all the anatomical structures involved in extension be preserved, especially the muscles and tendons?

2.7.3.2.1. Anatomy

These structures are summarized in Table 2.2. MTP extension is made possible by the extensor digitorum brevis (EDB) and the extensor digitorum longus (EDL), while the extension of the proximal (PIP) and distal (DIP) interphalangeal joints is provided by the lumbrical muscles. The plantar and dorsal interossei muscles expand on the extensor mechanism providing some tension but not active control. It should be noted that the EDB and the EDL do not have an expanding action on the PIP and DIP.

2.7.3.2.2. Biomechanics

In order to assess the importance of conserving the structures involved in extension, their physical mechanisms in human activity must be analyzed.

During anterior stance, there is a phase in which the pressure centers are level with and in front of the metatarsal heads (Figure 1.22). MTP extension occurs at this time.

What do the data tell us?

– Kinematics: kinematic data (expressed as links representing the limb segments) show that the tibia falls forward with the foot on the ground and the heel raised (Figure 2.33), which is corroborated by the dynamic analysis.

– Dynamics (Figure 1.22): the fall of the vertical force followed by a sagittal force peak attests to the fall in front of the body. This causes extension, which is therefore a passive phenomenon.

– Electromyography (EMG) (Figure 1.23): this passive phenomenon is confirmed by the fact that the electrical activity is at its lowest (anterior tibial and EDL tendons) when the MTP joints are extended.

2.7.3.2.3. Conclusion

Thus, we can conclude that MTP extension, which is specific to the human species and essential for the proper flex point, is a passive phenomenon that does not involve the muscles and tendons that are supposed to ensure it.

2.7.3.3. *Surgical consequences*

2.7.3.3.1. Selective sectioning of EDL

The deformities observed in claw toes are frequently accompanied by an upright position of the first phalanx related to MTP hyperextension (Figure 2.34). EDL retraction is consistent, creating a protruding dorsal cord visible under the skin (Figure 2.37). This is the longest tendon, with the greatest ability of storing elastic energy and, conversely, of retracting: this is why we choose to do a tenotomy here. This reduces the permanent extension of the MTP joint. Insofar as the EDL is not actively involved, it can be sacrificed without consequence, except in the fifth ray, as there is no EDB. Without remorse either, since this procedure does not hinder the mechanisms involved in the flex point. On the other hand, the associated transection of the EDB, the interossei – through the intermediary of their tendon expansions – and the lumbricals can lead to toe ptosis, which can hinder walking.

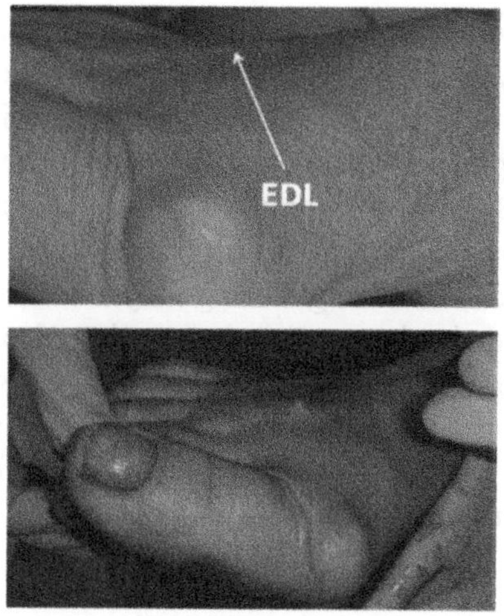

Figure 2.37. *Protrusion along a dorsal cord of the EDL (LEO) and site of the "suspended" tenotomy*

2.7.3.3.2. Technique (Figure 2.37)

The technical implications are highly specific: transection of the EDL (tenotomy) must be done in a "suspended" manner, taking advantage of the dorsal cord effect. It is essential to avoid placing the scalpel on the dorsal surface of the metatarsal to avoid damaging the EDB (*Extensor digitorum brevis* (Figure 2.38)) and avoid going into the intermetatarsal spaces to avoid the lumbrical and interossei muscles.

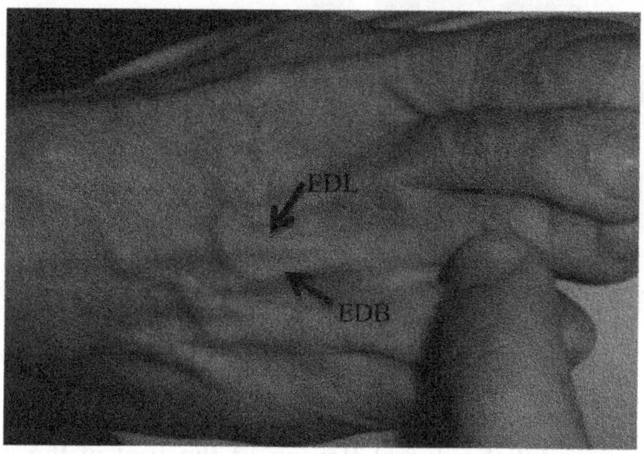

Figure 2.38. *Visualization of the extensor digitorum brevis (EDB) and longus (EDL)*

This transection should not be done by dorsal capsulotomy of the MTP joint, as it necessarily involves the associated transection of the EDL and EDB, which join at this level, and destabilizes the joint by opening it. The tenotomy should be performed two to three centimeters behind the MTP joint to avoid the EDB, but not further back as its effect could be nullified by the presence of a possible compensatory transverse tendon band coming from the extensor of the neighboring ray. Excessive tenotomies and capsulotomy may result in the toe ptosis mentioned previously. Procedures can be performed step by step, with increasing aggressiveness depending on the result obtained: if the transection of the EDL is insufficient, the EDB is then cut and a dorsal capsulotomy of the MTP is performed. It is important not to damage the interossei and lumbrical areas.

2.7.4. *Flexor tendons: which priorities?*

2.7.4.1. *Functional anatomy review*

In humans (Figure 2.33), gait is achieved by alternating plantar and dorsal flexions of the ankle and by the extension of the MTP joints. Flexion of the toes in the lateral rays stabilizes the forefoot on the ground and guides it towards the hallux, which is solely responsible for propulsion. Conversely, in apes (Figure 1.29), ankle dorsiflexion is permanent, gait function is in the midtarsal area and full, active flexion of all the toes is crucial for propulsion, grasping and arboreal locomotion (Figure 2.39).

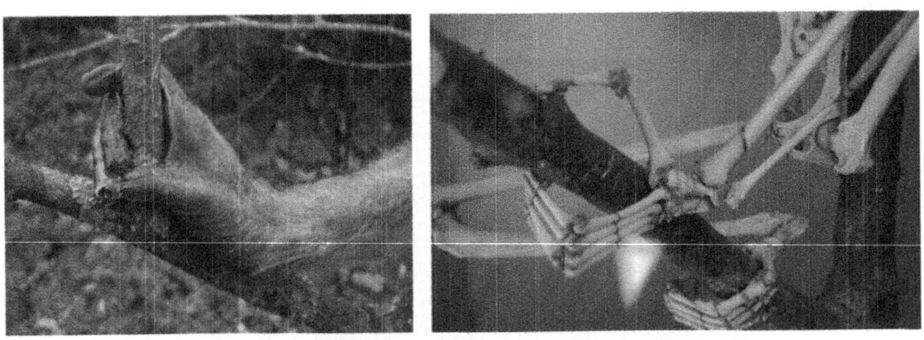

Figure 2.39. *Ape or lemur foot; grasping and arboreal locomotion*

Taking into account the dorsal flexion of the foot at the midtarsal level, flexion is essential at each segment for the foot to contact the ground.

Moreover, PIP flexion is essential for arboreal grasping (short flexor), whereas it is the overall rolling function of the long flexor that is dominant in humans.

	MTP	**PIP**
Extension	80°	80°
Flexion	30°	90°

Human

	MP	**PIP**
Extension	20°	0/20°
Flexion	110°	110/150°

Ape

Table 2.3. *Flexion and extension range of MTP and PIP joints in humans and apes*

The data in Table 2.3 show that the mobility sector is extension in humans and flexion in apes. Anatomically, this is reflected by the plantar orientation of the articular surfaces of the metatarsal heads – in flexion in apes, but in extension in humans – and the presence of retrocapital dorsal metatarsal stops blocking extension in apes (Figure 1.32a). The relatively large flexor muscle mass in apes illustrates its importance in contrast to humans. On the other hand (Table 2.3), humans have more important interossei and lumbrical muscles which ensure the flexion of the MP joints, and the winding action in the flexion of the PIP and DIP of the long flexor, by its distal insertion on the third phalanx.

2.7.4.2. *Possible anatomical sacrifices*

2.7.4.2.1. Tendons

Preserving the flexor digitorum longus (FDL) (Figure 2.40) seems necessary because it allows active plantar flexion and an effective plantar stance to be maintained, especially if the active flexion of the MTP joints is maintained by preserving the interossei and lumbrical muscles. This is the same conclusion that we reached at the end of section 2.8.3.2.3 on extensors.

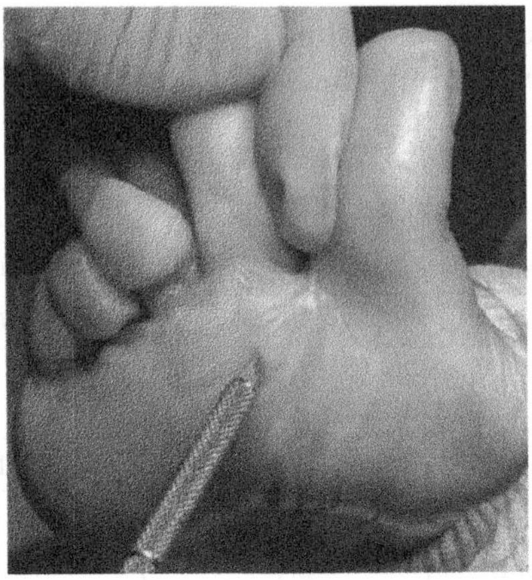

Figure 2.40. *Transection of the flexor digitorum longus (FDL) at the plantar base of P1 is not desirable for maintaining the toe curl function*

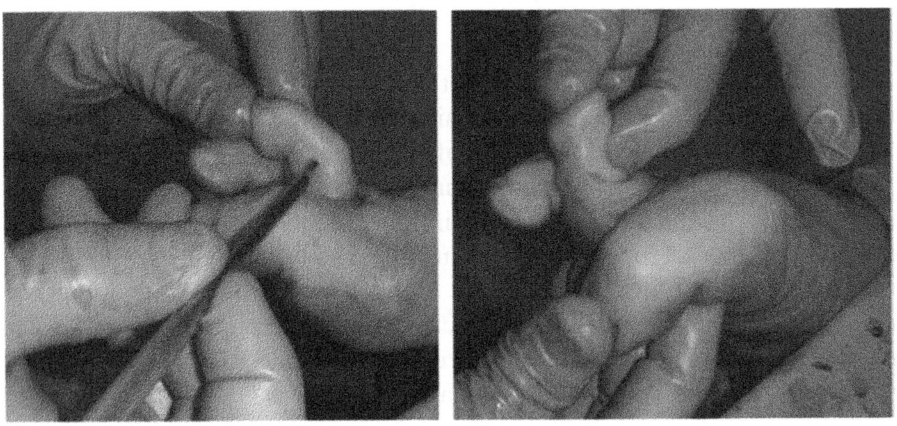

Figure 2.41. *Percutaneous transection of the flexor digitorum brevis (FDB) at the plantar surface of the PIP*

Under these conditions, neutralization of the PIP can be done without regret, either by tenotomy of the flexor digitorum brevis (FDB) (Figures 2.41 and 2.42) or by arthrodesis of the PIP of the toes (Figure 2.43). The latter procedure helps to correct a PIP flexion deformity and can be performed percutaneously.

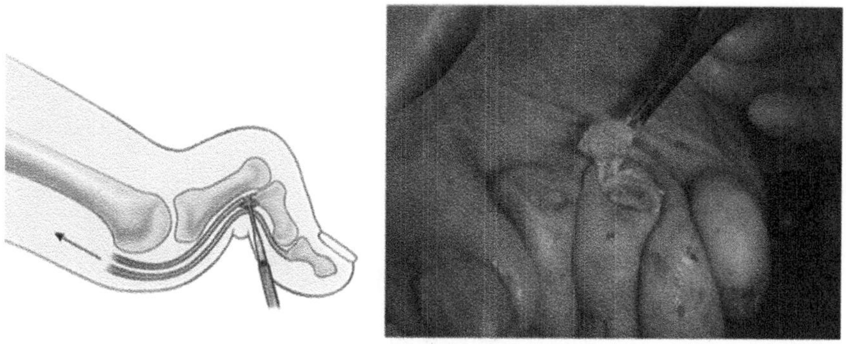

Figure 2.42. *View of the two FDB insertions at the base of P2, which are being tenotomized. For a color version of the figure, see www.iste.co.uk/cazeau/foot.zip*

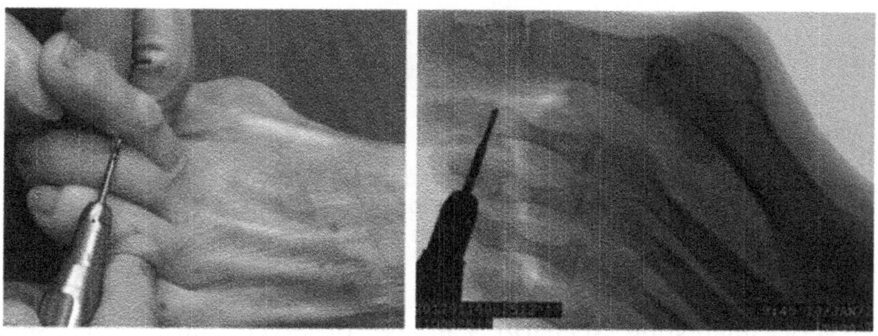

Figure 2.43. *Percutaneous PIP arthrodesis*

2.7.4.2.2. Bones and joints

Osteotomies can be performed to realign the first and second phalanxes, with the theoretical advantage of preserving joints and tendons. In practice, if the patient is not compliant with the post-operative instructions, follow-up and rehabilitation, there is a risk that adhesions between tendons and bony calluses in this narrow digital segment will have the opposite effect, with stiff toes "like a drum stick".

On the joints, resection arthroplasty or PIP or DIP arthrodesis can be performed. Regardless of which one is done, it is critical to preserve the action of proximal flexion (interosseous and lumbricals) and distal flexion (FDL).

2.7.5. *Unusual anastomose*

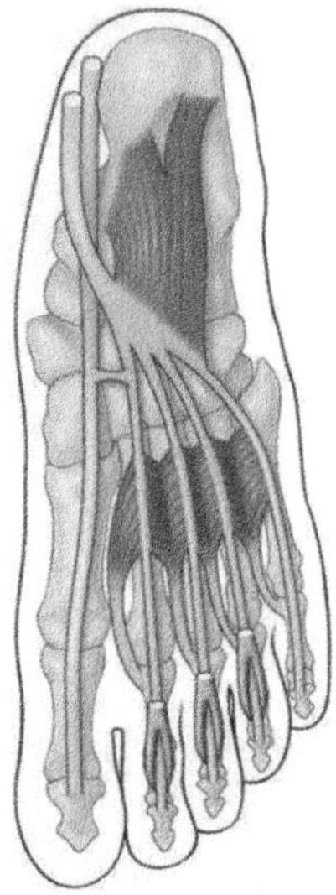

Figure 2.44. *Anastomoses between the hallux flexor proper (flexor hallux longus) and the common flexor of the toes (flexor digitorum longus) in humans. For a color version of the figure, see www.iste.co.uk/cazeau/foot.zip*

A very interesting scenario can be observed during the surgical shortening of M1. The presence of tendon anastomoses between the flexor hallux longus and the common toe flexors, which are inconstant in *Homo* species (Figure 2.44), makes them integral. The posterior displacement of one of them due to M1 shortening is responsible for traction, recoil of the toe flexor and the appearance of claw toes. This is called "syndrome d'attelage" in France.

This configuration in humans is an archaic trait, probably present in the common ancestor between humans and apes, because it is regularly present in ape feet, which have a very different redundant plantar interlocking of flexors for grasping (Figure 2.45). The treatment for claw toes consists of sectioning these tendon slips. This is typically a deviation from the original normality for the sake of utility.

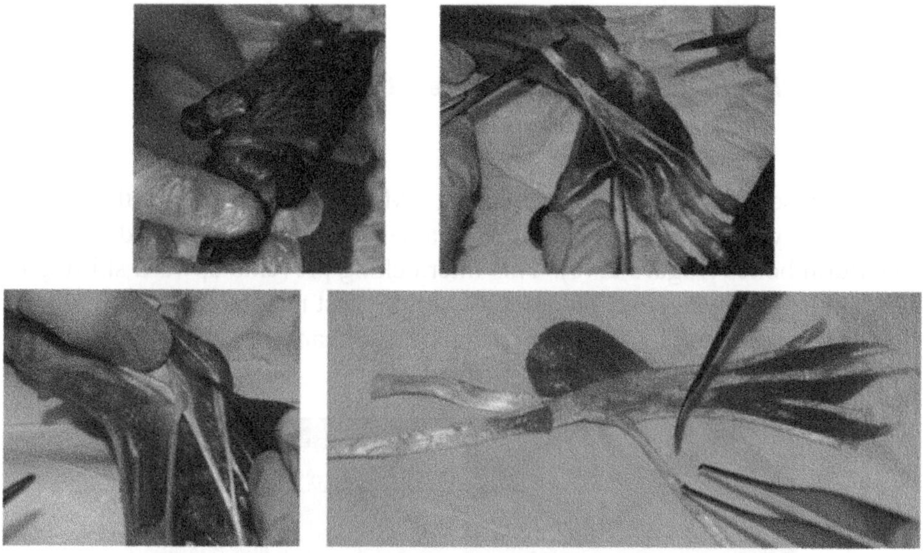

Figure 2.45. *Complex intricacy of the flexor hallux with the toe flexors in apes*

2.7.6. *Conclusion*

Claw toe surgery requires sacrifice.

Understanding of the particularities of human walking is inseparable from the analysis of its close relatives. It allows us to distinguish useful (function) from normal (anatomy) and, through this clarification, to sacrifice certain structures such as the extensor digitorum longus (except the fifth ray) and the FDB without remorse.

This also allows us to focus our attention on preserving other formations: the dorsal and plantar muscles, the extensor digitorum brevis (EDB) and the flexor digitorum longus (FDL).

Biomechanical analysis coupled with comparative anatomy work allows us to specify the relevant area of surgical action precisely as possible.

2.8. Second ray syndrome

2.8.1. *Medical and surgical considerations*

2.8.1.1. *Anatomy of the second ray*

The base of the second metatarsal (M2) is entrapped in the tarsus, articulating with the first and third metatarsals and the first, second and third cuneiform bones (Figure 2.46). This interlocking prevents upward shifting of M2 if hypertrophy occurs. The mobility of the first metatarsophalangeal joint (MTP1) is reduced by about 5°, much less than the 10° in the tarsal–metatarsal joints of the fourth and fifth rays.

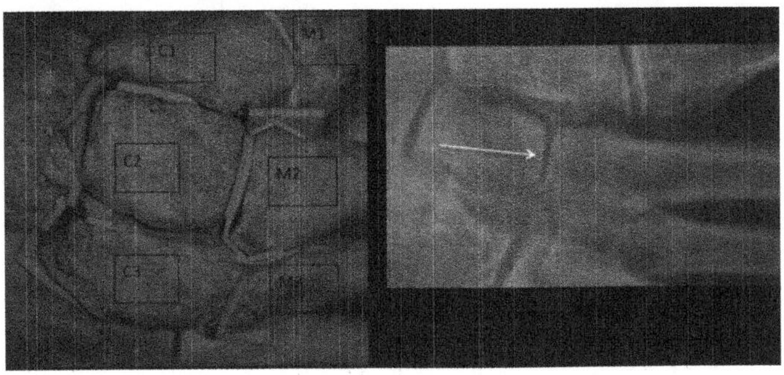

Figure 2.46. *Interlocking of M2 base. For a color version of the figure, see www.iste.co.uk/cazeau/foot.zip*

2.8.1.2. *Specific clinical entity*

By definition, this syndrome affects the second ray only, and is solely responsible for it. This eliminates all other causes of overloading, coming from the hind foot – such as pes cavus foot – or the forefoot. Excessive pressure can occur due to functional insufficiency of the first ray, caused by true or geometrical brevity of the first metatarsal, or acuneometatarsal laxity.

This is responsible for load transfer on the lateral rays, the most symptomatic expression of which will be on the second ray, as well as on the neighboring metatarsals. Under these conditions, symptoms such as metatarsalgia or hyperkeratosis diminish proportionally as we move away from the second ray. This is a departure from the strict definition of the second ray syndrome.

2.8.1.3. *Causes*

Before analyzing the causes, we must bear in mind that some human mechanical systems operate with the smallest of safety margins and that the slightest disturbance will lead to decompensation and the appearance of symptoms. The mechanical mode of operation and the anatomical environment explain the high frequency of pathologies, such as rotator cuff tears, compression of the median nerve at the wrist, patellar syndromes, ankle instability, etc. The second ray has its place on this list: the rigid entrapment of its base at the tarsus does not allow any compensation for excess length or obliquity of the M2 at the ground level.

2.8.1.4. *Clinical presentation*

The types of presentation are polymorphic, multiple and tiered, ranging from second toe forward to wedge-shaped backward. The following signs can be observed: claw deformity of the second toe, pain under the M2 head, plantar hyperkeratosis – especially if it is medallion-shaped (Figure 2.47) – alteration of the plantar plate generating instability that can lead to an irreducible dislocation.

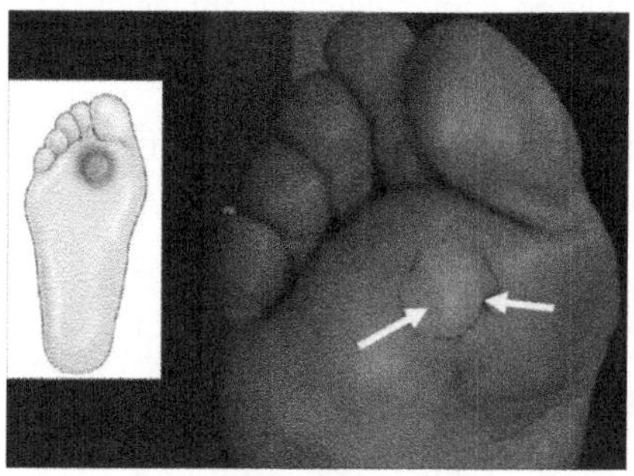

Figure 2.47. *Medallion hyperkeratosis under the M2 head*

The pain may also be related to a stress fracture or necrosis of the M2 head. There may be pain on the lateral sesamoid. Symptoms may manifest as a pseudotumor under the M2 head due to subcutaneous rupture of a bursa.

Certain types of stresses can manifest themselves as true fractures. These are the result of intense, repetitive stresses corresponding to a new type of physical activity. The lack of adaptation of the (otherwise healthy) bone to these situations leads to a fracture. It is as if the capacity for repair is exceeded by the intensity of the stress, which then manifests itself in microcracks.

Involvement of the M2 diaphysis may result in a stress (fatigue) fracture (Figure 2.48), sometimes recurrent, or in isolated signs of radiological overload (hyperostosis) manifested by cortical thickening (Figure 2.49).

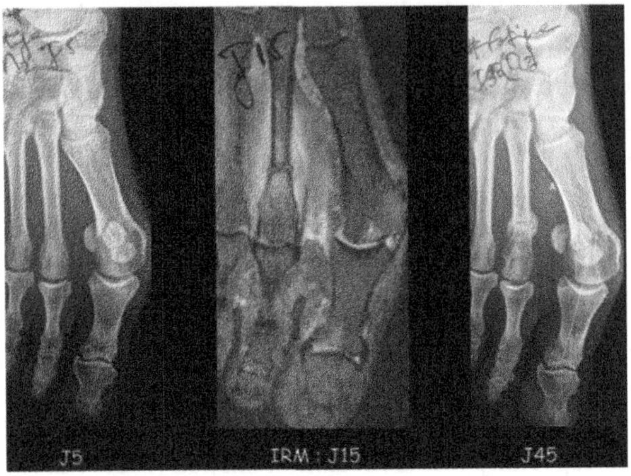

Figure 2.48. *Stress fracture of M2 invisible on D5 on standard radiographs, producing a large bony callus at D45; visible at D15 on MRI*

It is also possible to observe a stress fracture of the M2 base, a tarsometatarsal osteoarthritis (Figure 2.50).

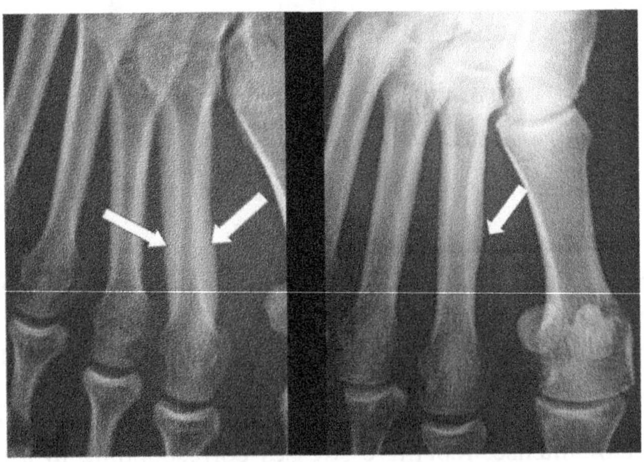

Figure 2.49. *Hyperostosis or thickening of the M2 cortices, a sign of overload*

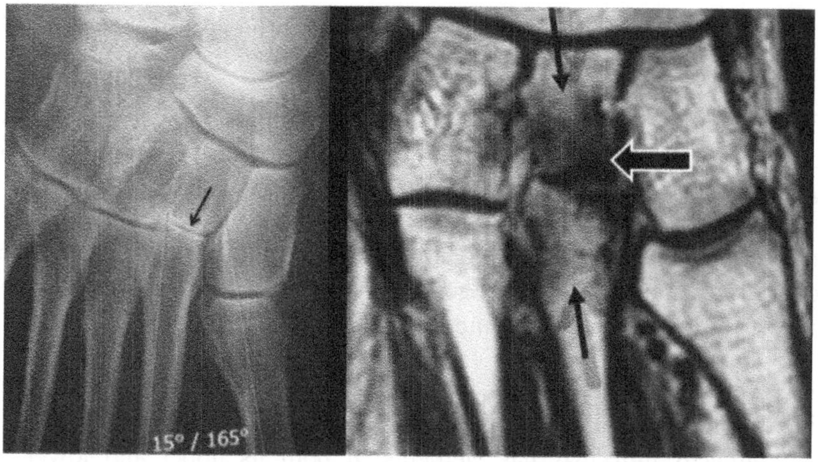

Figure 2.50. *Tarsometatarsal osteoarthritis, stress fracture of the second cuneiform and the M2 base*

Similarly, signs of overloading may be seen in the even more proximal segment and – in addition to pain in the second cuneiform – there is a fatigue fracture and/or necrosis visible on MRI (Figure 2.50).

2.8.2. *Second metatarsal in human primates, non-human primates and fossil hominids*

2.8.2.1. *Introduction*

The human foot is the result of several million years (Ma) of evolution, during which the role of the lower limb has changed somewhat due to changes in the environment since the first tetrapods emerged.

This has resulted in the appearance of a multitude of single or associated locomotor functions within a varied locomotor repertoire (quadruped, arboreal, occasional or permanent terrestrial biped, etc.), the result of a long process of cumulative adaptations by natural selection.

Other constraints have been created by lifestyles, physical activity or culture. For example, footwear has definitely been worn for 15,000 years and probably for 40,000 years (Kuttruff *et al.* 1998; Trinkaus 2005; Trinkaus and Shang 2008), and therefore appeared much later than the modern foot.

Consequently, understanding the current human foot requires looking at our closest primate cousins and how hominids have evolved over the last 7 Ma.

Interest in hominins appeared at the end of the 19th Century with the discovery by Dutch physician Eugène Dubois of the *Pithecanthropus erectus* from Java (Indonesia), literally the standing ape-man.

However, the fossil that has more recently influenced both scientific and media commentary is the famous Lucy, discovered in Ethiopia in 1974. This *Australopithecus afarensis* has elements of both its upper and lower limbs preserved, which has led to an ongoing debate about its locomotion (for more information on the chronology and bipedalism of fossil hominids, see Table 2.4).

Two years later, in 1976, Laetoli's remarkable footprints were unearthed in Tanzania and proved once again the existence of a very ancient practice of bipedalism with a *human* component through the footsteps of children and adults recorded in volcanic tuffs. Very recently, in Happisburgh, England (Ashton *et al.* 2014), remnants dating back a million years were found near an estuary. Rich in anthropological information, fossil human remains are all the more valuable because of their relative scarcity. Fossil foot remains have been extremely rare since the first hominid about 7 Ma ago. Only seven whole metatarsals have been collected between 7 Ma and 1.7 Ma, i.e. the period covering the establishment of so-called "human" bipedalism. Within those, only two whole M2 exist. The first, from South Africa and named Stw89 (DeSilva *et al.* 2012), was attributed to *Australopithecus africanus* based on dating between 2 Ma and 2.6 Ma (Pickering and Kramers 2010). The second, BRT-VP-2/73 (Haile-Selassie *et al.* 2012), dated 3.4 Ma, is an undetermined species from Ethiopia. In addition, there are five other M2 fragments belonging to Australopithecus (*Au. afarensis*, *Au. africanus*) and a probable *Homo habilis* (OH8).

Dating	Species	Locomotor mode	Indications of bipedalism	References
6–7 Ma	*Sahelanthropus tchadensis*	Biped? Hominid? absence of post-cranial elements	Discussion on the position of the occipital hole	Brunet *et al.* 2002, 2005; Wolpoff *et al.* 2002, 2006; Zollikofer *et al.* 2005
6 Ma	*Orrorin tugenensis*	Bipedal and arboreal/debated bipedal/absence of foot elements but presence of post-cranial elements	Cortical distribution of femoral neck, human type with the upper part being less thick	Senut *et al.* 2001; Pickford *et al.* 2002; Galik *et al.* 2004; Richmond and Jungers 2008; Puymerail 2011
4.5–5.5 Ma	*Ardipithecus ramidus*	Arboreal and bipedal more primitive than *Australopithecus*/presence of foot elements and various post-cranial elements	Cranio-caudal shortening of the ilium and strengthening of the spine's lordotic curvature	Lovejoy *et al.* 2009a, 2009b, 2009c, 2009d; White *et al.* 2009
3.8–4.2 Ma	*Australoputhecus anamensis*	Arboreal and bipedal/absence of foot but presence of tibia and femur	Talar surface perpendicular to the axis of tibial diaphysis	Leaky *et al.* 1995; Ward *et al.* 1999, 2001
3–4.1 Ma	*Australopithecus afarensis*	Arboreal and bipedal/frequency and type of bipedality debated/presence of foot elements and many different post-cranial remains	Short, broad pelvis, obliquity of the femoral shaft (adducted knee), ellipticity of the lateral femoral condyle, concavity of trochlea and protrusion of the lateral lip, human talocrural joint (movement of the leg in the sagittal plane with the foot), broad tuberosity on calcaneus	Senut and Tardieu 1985; Lovejoy 1988; Lovejoy *et al.* 2002; reviewed in Kimbel and Delezene 2009; Haile-Selassie *et al.* 2010

Dating	Species	Locomotor mode	Indications of bipedalism	References
Approx. 3 Ma	*Australopithecus africanus*	Arboreal and bipedal/more arboreal than *Au. afarensis*/debate about the human character of hallux and ankle/presence of foot elements and various post-cranial remains	Oblique femoral diaphysis, concave femoral trochlea with lateral lip protrusion, short, broad pelvis	Lovejoy and Heiple 1970; Heiple and Lovejoy 1971; Tardieu 1983; McHenry and Berger 1998a; Green *et al.* 2007
2.5–1.4 Ma	*Paranthropus*	Biped? Arboricultural? few post-cranial remains/lack of trust in the attribution of certain fossil parts between *Paranthropus* and *H. erectus* or even with chimpanzees	Femoral diaphyseal morphology resembles *H. erectus*	Aiello *et al.* 1999; Aiello and Dean 2002; McHenry *et al.* 2007; Dominguez-Rodriguo *et al.* 2013
1.9 million	*Australopithecus sediba*	Biped and arboreal/debate on whether it belongs to the genus *Homo* or *Australopithecus*	Femoral trochlea concave, lateral lip extremely prominent, pelvis short and broad with wings of ilium more vertical than *Au. afarensis*.	Balter 2010; Berger *et al.* 2010; Berger 2013; Churchill *et al.* 2013; DeSilva *et al.* 2013

Table 2.4. *Locomotor mode in non-Homo hominids and indications of non-foot bipedalism (Tony Chevalier)*

2.8.2.2. *Functional anatomy comparison between human and non-human foot*

It is interesting to note that in the genus *Homo*, the exclusive, specific and unique appearance of the second metatarsal base is clinically responsible for the second ray syndrome by locking the midtarsal area. By shifting the zone of mobility downstream, this suppresses the midtarsal flex point in non-human primates.

Extension at the metatarsophalangeal joints is therefore a human specificity, and conversely the flexion of the metatarsophalangeal joints is more adapted to grasping for arboreal movements. In the great ape, the flex point is shifted backwards to the midtarsal break.

This extended mobility is accompanied by locking of the extension of the metatarsophalangeal joints, related, on the one hand, to the plantar orientation of the cartilaginous surface of the metatarsal heads and, on the other hand, to the presence of a dorsal bone stop at the heads preventing the phalanges from lifting (Figure 1.32a).

One's "normality" is therefore not the same as the other's, even if both are lumped into the fuzzy functional anatomy concept of bipedalism. This anatomical criterion is fundamental in the study of human phylogeny.

A functional anatomy comparison of the human bipedalism model with that of the ape allows us to understand how the second ray is protected in the ape, and conversely to assess the degree of exposure in humans.

In describing the system from proximal to distal, the observations are as follows.

With regard to gait development, dorsal flexion of the ankle is observed in humans between the 60th and 88th percentiles of the stance phase (phase C), particularly when the ankle is at a right angle (point X) (Figure 2.33). The reverse is true in apes, which move with the ankle permanently dorsiflexed (Figure 1.29). This position prevents the efficient functioning of the bi-articular muscles observed in humans. This is all the more true since the muscle bodies of the posterior compartment of the ape descend to the calcaneus, without individualization of a calcaneal tendon, thus preventing the storage of elastic energy.

Gait development in humans is characterized by a calcaneal heel strike in valgus, a plantigrade phase initially in varus and then flat foot, the forefoot being placed in pronation to transfer the gait progressively towards the first ray, with the hallux having the very last contact with the ground. Just before this last phase, body weight is transferred under the M2 head which serves as a rotation point because of its fixed nature. In apes (Figure 1.30), the reverse is observed: the taligrade phase is in varus, due to rotation of the triangular shaped talus and the absence of a lateral calcaneal tuberosity. Then, the plantigrade phase takes place in varus, resting on the lateral edge of the foot, with gait development occurring by dorsal breakage of the midtarsal area. This occurs by the digitigrade phase, toes flexed. The mechanical forces are not located under the metatarsal heads, since the forefoot is flat, horizontal, due to the lack of a longitudinal arch. There is no involvement of the hallux during the propulsion phase in the chimpanzee.

The presence of this longitudinal arch in humans increases the integration effect of M2; the bone makes an angle of about 15° with the ground, the anterior face of the second cuneiform being angled downwards and forwards. It is the opposite for ape feet, with M2 being horizontal.

Of note, the study of the comparative strength of the metatarsals paradoxically shows that the second metatarsal is the weakest of all metatarsals in humans. This may be the reason for the frequency and polymorphism of the second ray syndrome.

2.8.2.3. *Robustness of the metatarsals, especially the second*

In 1974, a comprehensive analysis of metatarsal robustness was presented by Riesenfeld. The mean signature 1-5-4-3-2, which designates the order of the five metatarsals from the most robust to the least robust, is unique in humans. Among the other great apes, four have M5 as the most slender of the metatarsal group (*Hylobates*, *Symphalangus*, *Pan*, *Gorilla*), while *Pongo*'s M5 is the third most robust metatarsal. Riesenfeld (1974) mainly focused on M1 and M5. He concluded that his data link the 1 + 5 robustness model (M1 being the most robust and M5 the second most robust) with terrestrial displacement, and the 1 → 5 model (M1 being the most robust and M5 the least robust of the metatarsals) with arboreal displacement. Robustness would be more an indicator of substrate than of a particular type of locomotion. As for M2, which is the main focus of this chapter, among the six genera of great apes considered, it is the least robust in *Homo* and *Pongo* (brachiator/suspensor). Of the 51 primate genera studied by Riesenfeld (1974), M5 is generally the least robust. Those in which M2 is the most slender, like in humans, do not belong to a particular locomotor group.

2.8.2.4 *Indentation or integration of the second metatarsal*

Integration of the second metatarsal can be understood by visual observation without measurements relative to the other metatarsals, by measuring the distance between the proximal joints of the medial (C1) and intermediate (C2) cuneiform, or by the angle between the metatarsal axis and the ground.

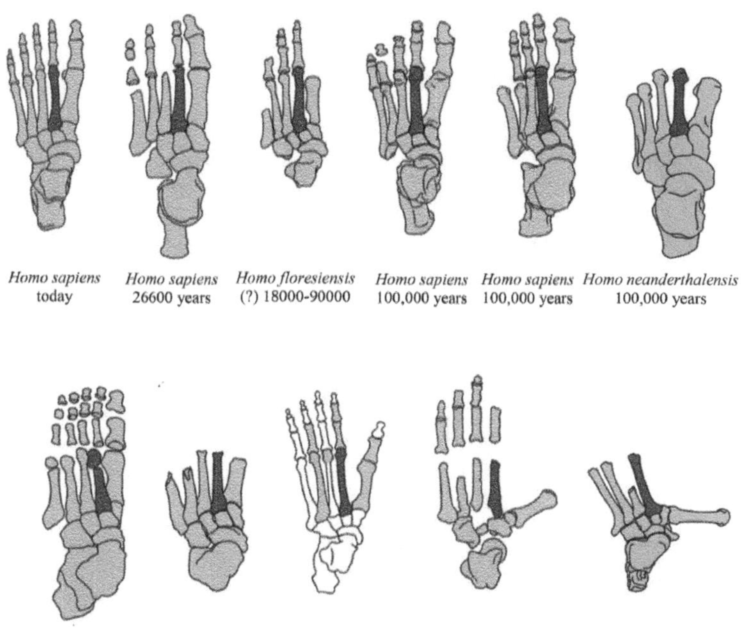

Figure 2.51. *Feet of current hominids and fossils plus Oreopithecus (drawings made by Christelle Milizia from illustrations published in Vandermeersch 1977; Kölher and Moyà-Solà 1997; Aiello and Dean 2002; Trinkaus and Svoboda 2006; Jungers et al. 2009; Pontzer et al. 2010; Susman et al. 2011; Haile-Selassie et al. 2012; Pablos 2013) (thanks to Tony Chevalier)*

The second metatarsal is integrated in humans due to a reduction in the anteroposterior–posterior length of the intermediate cuneiform (C2), which places it in contact with the three cuneiforms. It is a cornerstone within the transverse arch. The drawings presented in Morton (1924), Fleagle (1999), and Aiello and Dean (2002) (Figure 2.51) show a varied set of primate feet, some of which are reused in this work (Figures 2.51 and 2.52).

In this figure, humans cannot be differentiated based on the embedding criterion, except for very distant forms that clearly have a non-integrated M2: gibbon, macaque, tarsier, *Nycticebus* (Figure 2.52). However, Morton's (1924) illustrations, not shown here, do not show this as clearly. On the other hand, integration also exists in various primates such as *Pan* and *Gorilla* (Figure 2.52).

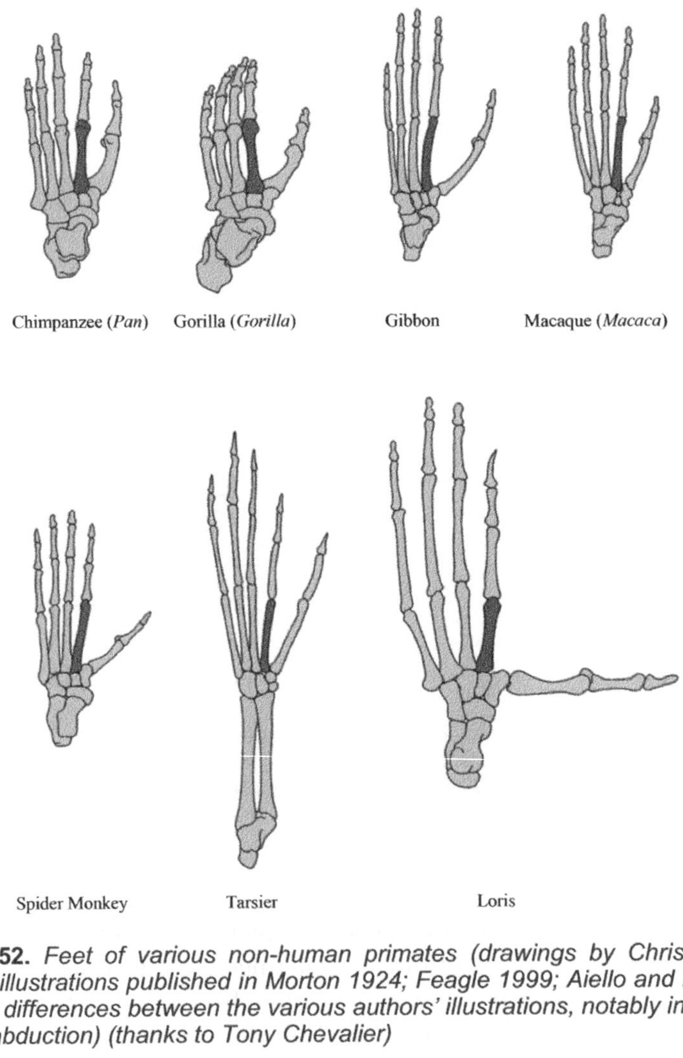

Figure 2.52. *Feet of various non-human primates (drawings by Christelle Milizia based on illustrations published in Morton 1924; Feagle 1999; Aiello and Dean 2002. There are differences between the various authors' illustrations, notably in the degree of hallux abduction) (thanks to Tony Chevalier)*

It appears that primates with really distinct locomotion have this M2 integration. Consequently, *its origin was prior to* the appearance of permanent bipedalism in *Homo* species. However, in all non-human cases, we can consider that hallux abduction partially releases the second metatarsal. In addition, another indirect sign of the limited mobility of the M2 is the presence of an articular facet at the lateral face of the M2 base, clearly delimited or blunt, resulting from contact with M1.

In humans, the increase in contact between C1 and M2, i.e. the proximity of the anterior joints of C1 and C2, was evaluated by Schultz (1930) by relating this distance to the length of the foot. He notes that for humans this is 6.4%, for mountain gorillas 5%, for lowland gorillas 1.3% and for chimpanzees 1–2%. This morphology is an adaptation for foot stability during terrestrial locomotion (Aiello and Dean 2002). In fact, mountain gorillas (*Gorilla beringei beringei*) are the most terrestrial non-human hominoid adult primates (Schaller 1963; Tuttle and Watts 1985; Doran 1996, 1997; Remis 1998), and here, like humans, they have the highest relative distance between C1 and C2.

2.8.2.5. *Metatarsal-ground angle*

It turns out that the orientation of the second metatarsal relative to the ground, relative to the longitudinal arch of the foot and the arrangement of the phalanges (flat on the ground at rest), causes direct contact between the M2 head and the ground, stimulating the integration (in the sense of locking) of the M2 base in its tarsal space.

2.8.2.6. *Consequences*

These three observations denote the originality of M2 integration in humans. The result of M2's anatomical position is extremely restricted mobility when the foot is put into action. M2 represents the fixed longitudinal axis around which the foot can move (rotate) during supination and pronation movements of the forefoot (Kelikian 1965; Huson 1991).

Based on the study of non-human primate morphology and the relative abundance of foot data over the course of human evolution, it seems consistent to assume that in fossil hominids, the second metatarsal was embedded in the tarsal row. The foot of *Homo habilis* (OH8) confirms this. However, given the above comments, we should note that there are different types and degrees of integration.

2.8.2.7. *Metatarsal torsion and foot arches*

Humans have a high transverse arch and an even more pronounced longitudinal arch on the medial side, the whole forming a remarkable arch of the foot (Figure 2.53). The transverse arch is also observed, to a lesser degree, in chimpanzees and other apes, while the longitudinal is absent (e.g. Weidenreich 1923; Morton 1935; Rose 1994; Sarmiento 1994) (Figure 2.53).

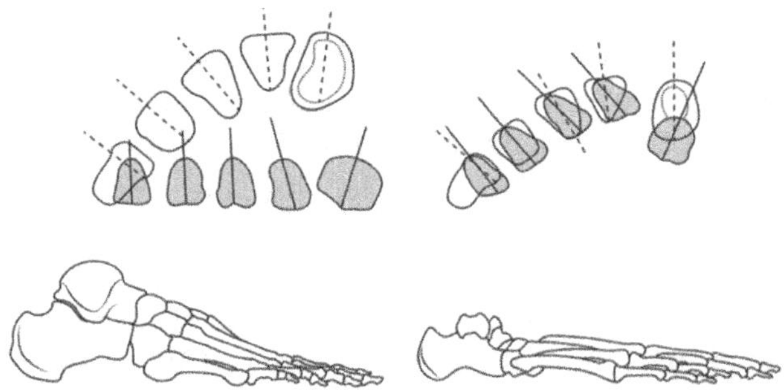

Figure 2.53. *Cross-sections through the base and head of human metatarsals on the left and the gorilla on the right; note the transverse and longitudinal arch in humans (thanks to Tony Chevalier)*

The pressure experienced by the metatarsals is directly related to the foot's morphology, and in particular to the existence of a single or double arch and various degrees of development. The metatarsal heads are particularly subject to high plantar pressure in primates during their movement (Wunderlich 1999). Therefore, as Drapeau and Harmon (2013) state, the study of metatarsal orientation is interesting because it reflects the positioning of the toes and their direction in flexion and extension. This is important in determining how the foot is used during locomotion.

Torsion of the metatarsal heads relative to their base is a means of specifying their orientation and appears very interesting for fossils when isolated elements are unearthed. The measurement of the metatarsal torsion corresponds to the angle between the axis of the head and the base of the metatarsus; a positive angle corresponds to the eversion of the head relative to the base and a negative angle to the inversion of the head relative to the base. In other words, relative to the base of the metatarsus, in inversion, the dorsal face of the head tilts laterally and the plantar face approaches the medial edge of the foot (Figure 2.54).

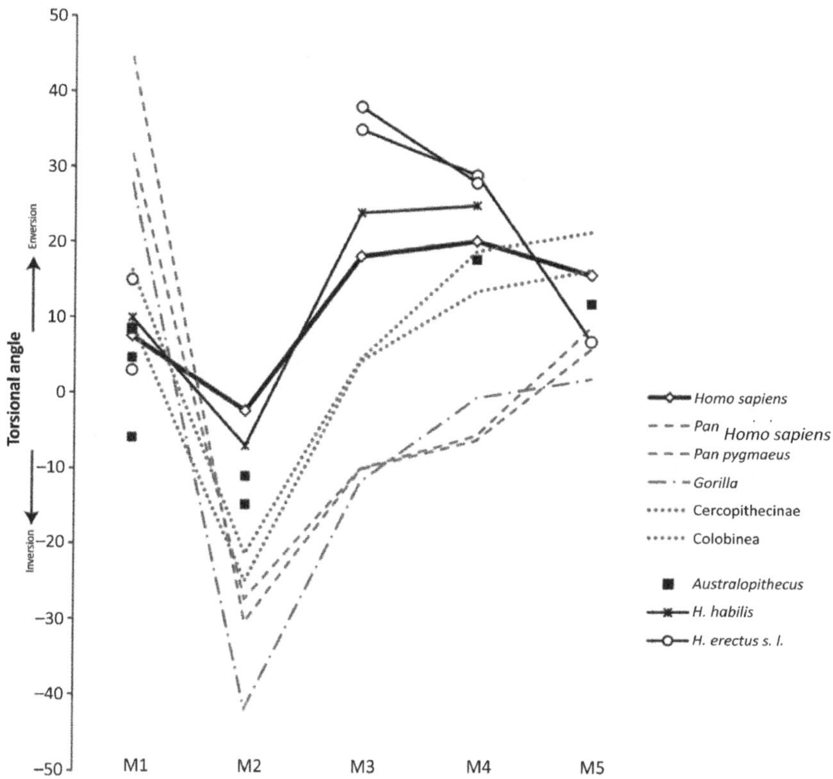

Figure 2.54. *Metatarsal torsion (graph based on the data from Pontzer et al. 2010; Drapeau and Harmon 2013). Different species of Australopitheca make up this group.* Homo habilis *refers to the specimen OH8.* Homo erectus sensu lato *refers to the specimens of Dmanisi (archaic representatives of this taxon) (thanks to Tony Chevalier)*

In humans, a slight torsion of the M2 and M1 heads, or even the absence of torsion, keeps the plantar face of the head in contact with the ground. The lateral foot is everted.

On the contrary, in great apes (*Pan*, *Gorilla*, *Pongo*) as well as in Cercopithecinae monkeys (DeSilva *et al.* 2012; Drapeau and Harmon 2013), internal torsion is very significant and places the M2 and M3 heads opposite to the first metatarsal.

In summary, humans have a small M1 and M2 torsion angle and a sharp eversion from M3 to M5.

These data indicate the uniqueness of M2 in humans. M2 torsion is an unambiguous sign of an ability to grasp with the foot since it is in inversion in the prehensile groups and without torsion in permanent bipedal humans.

2.8.3. *Treatment*

Treatment should focus on restoring function, i.e. painless metatarsal support at the transition from gait function to the end of the stance phase.

If the second ray is too long, it should be shortened (Figure 2.24). This shortening automatically and geometrically raises the head, thus also contributes to reducing head pressure.

When the second ray is normal in length, an isolated raising osteotomy should be performed without shortening to avoid unbalancing the forefoot. This can be achieved by an osteotomy at the M2 base, a dorsal basal metatarsal subtraction osteotomy (Figure 2.55).

Resection of a bony section with a dorsal base and plantar metaphyseal hinge geometrically provides a large leverage effect and significant head ascension. The gait can be considered normative and it restores the organ's "normal" anatomy.

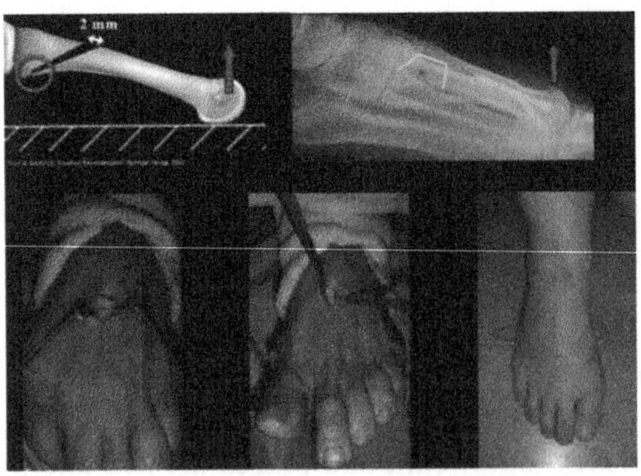

Figure 2.55. *Osteotomy to raise the M2 base by removing a section of the bone at the dorsal base. For a color version of the figure, see www.iste.co.uk/cazeau/foot.zip*

2.8.4. *Conclusion*

Second ray syndrome has the particularity of being able to manifest itself in a polymorphic manner along its entire length, from the second toe to the second cuneiform. It is a pathology specific to the genus *Homo*, a consequence of midtarsal locking. Comparative analysis of human and non-human primates and hominids shows that this is an innovation in the context of human bipedalism.

2.9. Morton's neuroma

2.9.1. *Introduction*

"Morton's neuroma" is one of the classic causes of compressive metatarsalgia. In addition to clinical examination, diagnosis often requires additional tests (ultrasound) to confirm the presence of swelling, as well as to rule out the many differential diagnoses. As always, surgery is reserved for the failure of non-operative[7] treatments.

2.9.2. *Pathological anatomy review*

The interdigital nerve (Figure 2.56) travels in a narrow path in the intermetatarsal space below the deep intermetatarsal ligament (DIL).

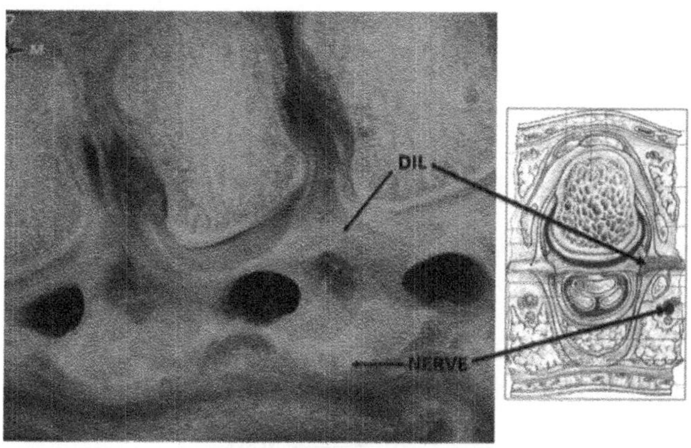

Figure 2.56. *Frontal section showing the deep intermetatarsal ligament (DIL) in green and the digital nerve in black. For a color version of the figure, see www.iste.co.uk/cazeau/foot.zip*

7 https://www.chirurgie-du-pied.fr.

With compression, repeated microtrauma leads to nerve ischemia, with degeneration responsible for its fusiform deformation (Figure 2.57), incorrectly called neuroma. This is not a nerve tumor; it is the pathological transformation of a normal nerve.

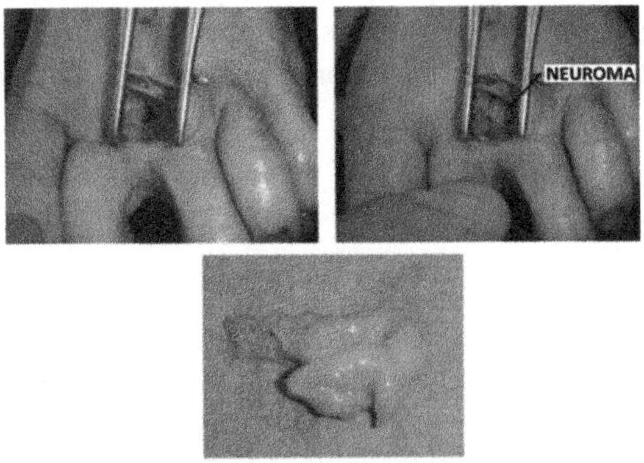

Figure 2.57. *Intraoperative view with the surgeon pushing upwards into the intermetatarsal space to show the neuroma, which was then removed. For a color version of the figure, see www.iste.co.uk/cazeau/foot.zip*

The anatomical environment is shown in Figure 2.56. Impingement primarily occurs by shearing of the nerve at the anterior edge of the DIL (Figure 2.58).

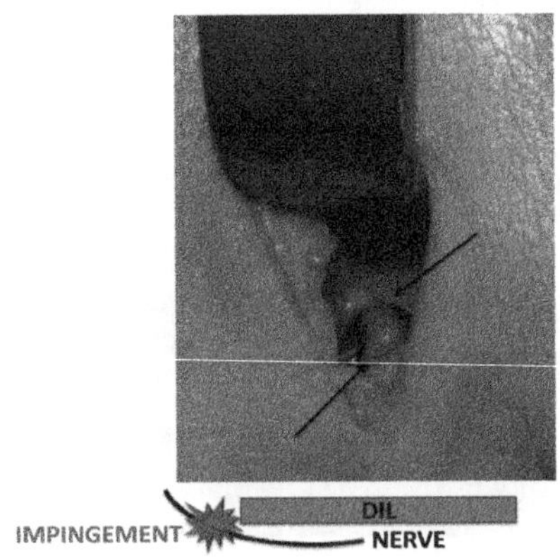

Figure 2.58. *Nerve being sheared at the anterior edge of the DIL. For a color version of the figure, see www.iste.co.uk/cazeau/foot.zip*

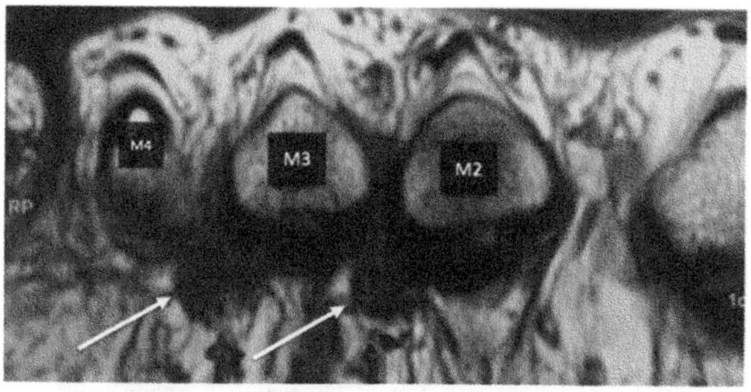

Figure 2.59. *Frontal MRI section showing two neuromas with plantar development*

In these conditions, the pathogenic factors are known; the stance phase pushes the neuroma between the metatarsal heads, tight shoes trap the neuroma and MTP hyperextension situations, especially equinus. As the neuroma is developing on the plantar side, as shown in the MRI frontal section (Figure 2.59), the stance phase pushes the neuroma towards the DIL (Figure 2.58). Tight shoes reduce the intermetatarsal space. Any MTP extension, especially equinus deformity due to wearing of high-heel shoes, pushes the neuroma back against the anterior edge of the DIL because the DIL is above it (Figures 2.56, 2.58 and 2.59).

2.9.3. *Clinical consequence*

Typically, metatarsalgia-type pain occurs with walking and increases with prolonged walking. At its extreme, patients say it is so intense they feel the urge to take off their shoes, which then provides relief. Wearing high heels and tight shoes aggravates the symptoms.

2.9.4. *Surgical options*

The neuroma cannot be removed (Figure 2.57) without severing the nerve because the nerve itself is deformed. The treatment is radical, effective and the resulting zone of anesthetic is not very disabling, as it is located in the interdigital space. Recurrences can only occur with the appearance of a neuroma, this time by removal.

Surgical decompression without excision is justified by the pathological compression of the interdigital nerve by the deep intermetatarsal ligament (DIL). Surgery then consists of pure decompression by transecting the ligament. This can be done percutaneously (Figure 2.60). This treatment is like the physiology but leads to more painful recurrences. There is no area of anesthesia.

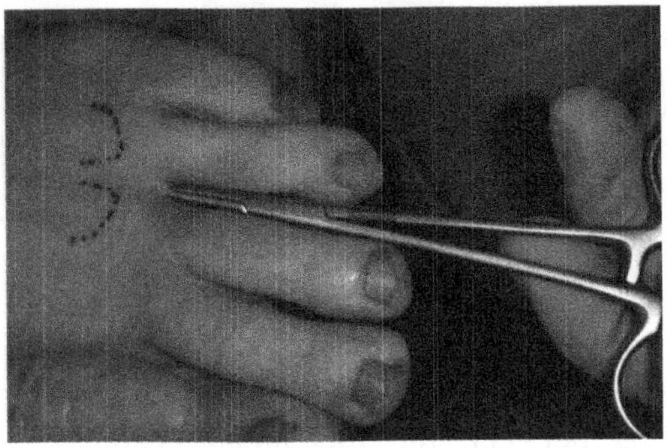

Figure 2.60. *Decompression without percutaneous excision*

2.9.5. *Choosing between utility and normality*

Nerve removal is a significant departure from normality. This is a rare medical situation, as it is usually possible to remove a true nerve tumor while preserving the nerve. It is nevertheless useful because, in the absence of a nerve, the pain disappears. Isolated decompression is closer to normality, but the long-term results are less reliable.

2.10. Appendix: anatomy and function of tendons

Dorsal interossei muscles

Occupy the first four intermetatarsal spaces. Inserted on the lateral +/- plantar faces of the posterior part of the first phalanx (P1) of T2 (DIO1 and DIO2), T3 (DIO3), T4 (DIO4) and expansion (unimportant) to the extensor tendon.

ACTION – MTP bending, P/R spacing at the second ray, PIP DIP extension

Extensor digitorum brevis muscle

Four muscle bodies coming from the calcaneum, oblique forward and inward +++. Four skeletal tendons under the EDB tendons which are overlapping. The first goes *under* the proper extensor tendon from the big toe to the dorsal and the lateral part of the base of P1, *not* attached.

The second, third and fourth tendons attach and merge with the outer edge of the corresponding ECO tendon at the MP level.

Angle between the CE and LE and the surface position allows the selective cutting of the EDL. Note: no EDB in the fifth ray (rare expansion at LE).

ACTION – MP extension and lateral inclination

Extensor digitorum longus muscle

Four tendons from T2 to T5, with the fifth constituting a true clean extensor, unique *attachment*: a few fibers at the lateral edges of P1 and a median strip on the dorsal side of the P2 base.

Two lateral slips join at the P2 head to insert at the P3 base.

Each tendon receives a homologous tendon of the EDB, dorsal expansion of the lumbrical and interossei and perforating fibers of the plantar aponeurosis.

ACTION – Powerful MP extension, no action on PIP and DIP++

Flexor digitorum brevis muscle

Superficial plane of the plantar region, superficial relative to FDL.

Four tendons dividing after MP1 at P1 into two slips that insert on the lateral edges of the underside of P2 of T2, T3, T4 and T5.

ACTION – Bending of PIPs

Flexor digitorum longus muscle

Before its division, sends an expansion to the flexor hallucis longus (*syndrome d'attelage*). Receives at its outer edge the quadrate muscle of Sylvius (accessory of the calcaneus to the FDL).

Four terminal tendons based on P3 of T2, T3, T4 and T5, perforating the FDB.

ACTION – Toe flexion (MP, PIP, DIP)

Lumbrical muscles

Belong to the deepest plane of the plantar region. Located in the spaces between the flexor tendons, in small fleshy adnexal components. Connected by a small tendon on the inner side of the base of the first phalanx, on the medial side of the MP and on the long extensor tendon of the toes by tendon expansion.

ACTION – Bending of MP, PIP and DIP extensions

Plantar interossei muscles

Three PIOs, on the same trajectory as the DIOs, emerge from M3, M4 and M5. A tendon at the lateral posterior end of P1, unimportant tendon expansion on the long extensor of the toes.

ACTION – Moving T3, T4 and T5 (third toe...) closer to the axis of the second ray MP bending, PIP DIP extension

3

Rapid Healing Through Immediate Use of the Operated Foot: A Purpose

3.1. Introduction

The purpose of any surgical procedure is to bring the patient as close as possible to recovery ("It is not the function of medicine to restore a patient to health, but only promote this end as far as possible", according to Aristotle, *The Art of Rhetoric*). For this, we must comply with the minimal specifications, such as implanting a hip prosthesis in case of osteoarthritis, and in the right position. This is indispensable, but our goals are not limited to this. Placing the organ attacked by the surgical procedure back in its usual working context as quickly as possible improves the odds of success. For a hip prosthesis, the mechanical forces during the stance phase trigger bone conduction inside the implant, and will help it to integrate properly into the body. Conversely, a bone immobilized by a metal plate that is too rigid and enclosed in a cast will deteriorate. This is the same problem faced by astronauts; they lose between 2% and 10% of their bone mass per month because they are no longer subject to the forces of gravity. An elderly person who has undergone hip surgery and is unable to resume walking within 48 hours of the operation will potentially have multiple complications that can lead to death if he or she is confined to bed. A foot operated on and suffering the same fate will have less dramatic but disabling complications with each step. Any unused organ atrophies; bones become demineralized and tendons lose their elastic properties.

Immediate resumption of the full weight bearing and immediate mobilization of the toes considerably reduces complications. This is how the concept of percutaneous and minimally invasive surgery should be defined.

3.2. The need for a concept

A concept is defined by the inseparability of a finite number of components. The concept of percutaneous surgery can be defined by the following inseparable components:

– geometric correction of the deformities in the three planes of space;

– reliable mechanical stability to allow immediate full weight bearing before any biological healing has occurred;

– minimal surgical damage to the anatomical structures being operated on;

– anesthesia technique that will prevent the painful postoperative storm associated with conventional techniques.

The benefit is faster functional recovery, which equates to a better final result.

Since they are inseparable components, they must be held together: they form a system between them and characterize percutaneous surgery in its specific, radically different concept.

3.3. New percutaneous and minimally invasive techniques

To better understand what these new techniques represent, they must be placed in their historical context.

3.3.1. *Development of percutaneous surgery*

Over the last 20 years, forefoot surgery has changed a lot in France. The technical procedures, which initially targeted tendons, ligaments and joint envelopes, have been extended to the bone itself. Before 1980, operations were mainly performed on these soft tissues, involving forcibly re-tensioning the distended elements located on the convexities, and in lengthening or sectioning what was retracted in the concavities. Pathophysiologically, the medial protrusion in hallux valgus is primarily related to the angulation of the first metatarsal (M1) (metatarsus varus) with the first phalanx of the hallux (P1). Attempting to forcefully reduce the bone axis by tensioning the soft tissues is doomed to fail. The logic of bone realignment then appeared, disseminated in France by Dr Patrice Diebold (Nancy) and Dr Samuel Barouk (Bordeaux).

Whether performing a diaphyseal, distal or proximal metaphyseal osteotomy, these techniques all target the bone itself, and thus require an understanding of the deformities and their correction in the three planes of space. A preoperative strategy can be established on the basis of precise biomechanical data, on the exact calculations of metatarsal lengths and on the size of the angular corrections to be made. Decomposing the phenomena to be corrected in the three spatial planes approaches reality and helps in understanding and selecting the corrections to be made.

3.3.2. *Principles of percutaneous surgery*

As in conventional surgery (Figure 3.1), percutaneous surgery consists of performing soft tissue procedures and osteotomies, but this time through a very small incision, with the help of *Beaver®* type scalpels and motorized burrs to cut the bone (Figure 3.2).

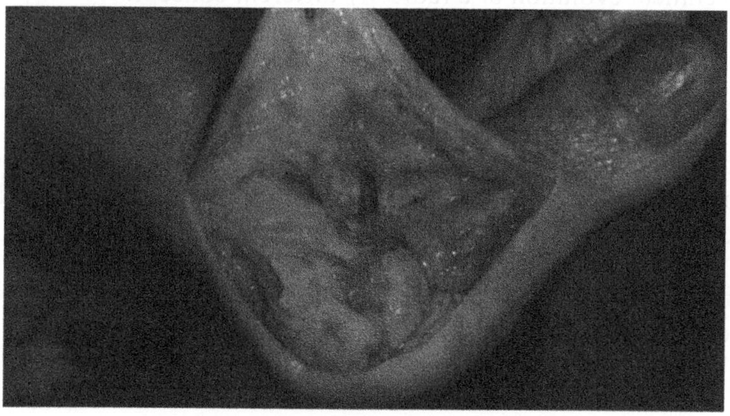

Figure 3.1. *Open approach to conventional surgery, aggressive tissue disruption*

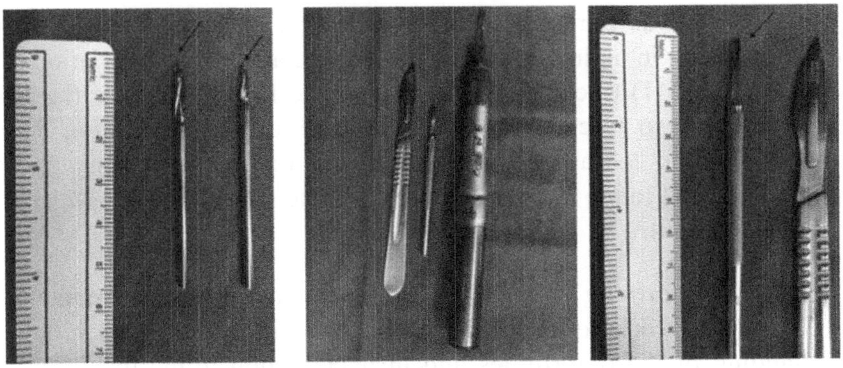

Figure 3.2. *Percutaneous surgical tools; motorized osteotomy burr and mini scalpel (Beaver®)*

The procedure is controlled during the operation by means of fluoroscopy. Percutaneous surgery is difficult to practice because there is no visual control other than fluoroscopy and because the culture of the procedure is very different from conventional surgery. Whether it represents a simple evolution of conventional surgery or a profoundly different concept will be evaluated later. It requires specific training from the experts who introduced percutaneous surgery of the forefoot in France.

This training is based on theoretical courses and cadaver workshops. Is it a simple technical evolution or a radically different concept?

The debate does not aim to contrast these new techniques to classical surgery, demonstrate the superiority of one over the other and maintain the myth of an agonizing struggle between the ancient and the modern. Dialogue is possible between the various proponents because there is a prior consensus: functional anatomy must be restored for the pain to disappear and for the physiological gait cycle to be resumed.

Minimally invasive surgery is gaining ground in all disciplines of orthopedic surgery, and the foot is no exception. The advantage for the patient lies in a theoretically faster functional recovery, for many reasons, not just because of small scars[1].

1 https://www.chirurgie-du-pied.fr.

3.3.3. *Biomechanical translation of this new conceptual approach*

Take the example of percutaneous osteotomies to raise the lateral rays in the case of pain under the forefoot due to excessive pressure.

To simplify, let us consider a system composed of two blocks, separated by the future distal metatarsal minimally invasive osteotomy *(DMMO)* (Figure 3.3a and b).

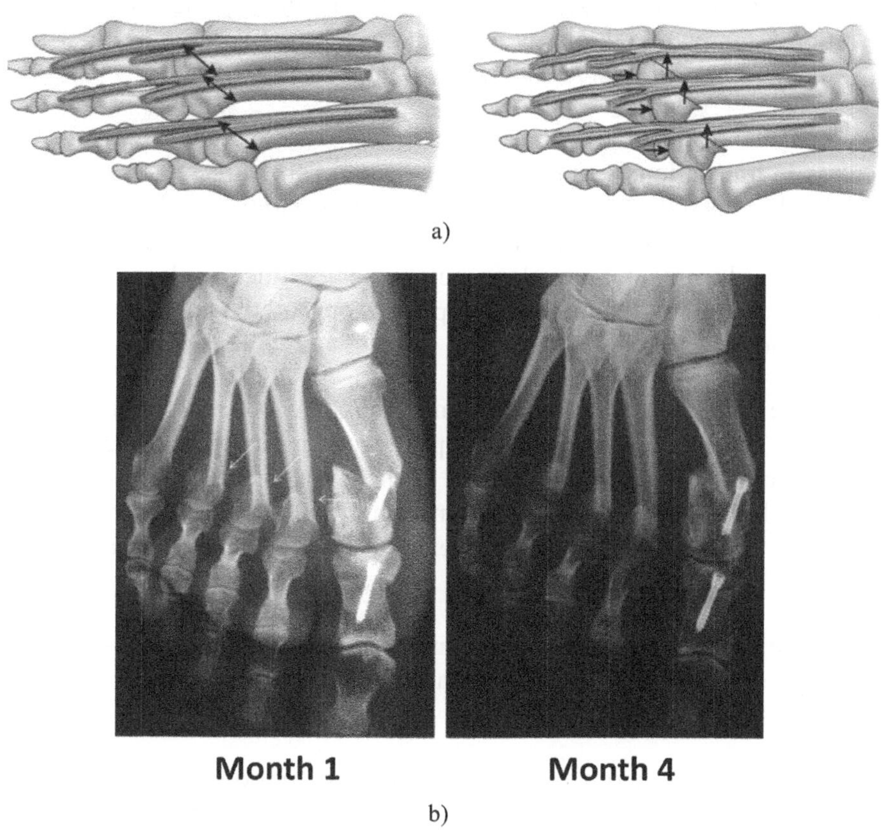

Figure 3.3. *a) Separation of two anatomical blocks using the DMMO technique. b) Bone callus development after DMMO between months 1 and 4. For a color version of the figure, see www.iste.co.uk/cazeau/foot.zip*

The proximal block consists of the posterior part of the forefoot extending forward to the osteotomy line. The second block, just below, consists of the neck and head of the metatarsal, the metatarsophalangeal joint and the toe in extension. The muscular tendinous and aponeurotic elements inserted beyond the future osteotomy line exert a resultant rearward traction force. Bone elements resist the same force in the opposite direction; these forces form an action-reaction pair according to Newton's third law.

These elements can be compared to a rope. The tension on this rope is the force exerted on the fixed elements located at its end materialized by the downstream osteo-articular block. The tension on this rope is an energy reserve that will be restored during the reverse posterior translation operation. This potential energy is related to the fact that the osteo-articular block below is placed in a tension force field.

When the resultant of the tension forces is zero, after osteotomy, the block is in translational equilibrium. The sum of the forces then acting on it is zero, because the automatic posterior translation of the heads after osteotomy eliminates the prior tension on the rope. This is the advantage of letting the adjustment take place automatically; the displacement of the block, according to the osteotomy plan, stops at the moment when the tension tends towards 0.

We are then in stable equilibrium, where the potential energy is at a minimum. The associated plantar support exerts an upward vertical force to ensure that the posterior displacement of the block is accompanied by ascent of the block. It is likely that the elimination of pain is related to the removal of tension forces, and also to the removal of plantar compression of the heads.

In most cases, we can see an alignment of the metatarsal heads with each other and often a slightly different Maestro parabolic curve on postoperative radiographs.

This principle of automatic correction is sometimes observed during an M1 distal osteotomy (chevron) performed for metatarsus varus; the automatic lateral displacement of the metatarsal head occurs as soon as the second cut is completed, without even exerting pressure on the head.

Classical preoperative planning, aimed at achieving the perfection of a normative model (Maestro's parabola), has proven empirically effective despite the approximations of dorsoplantar radiographic views, which does not discriminatively take into account the metatarsals' obliquity and length.

The static model of stable equilibrium through percutaneous surgery gives the same anatomical results, and immediate pain relief. Immediate complete plantar support by means of a rigid-soled shoe is not a suggestion, but a requirement as it is indispensable to the success of the technique and a direct extension of the surgical procedure.

This is a far cry from the heel support shoes (Figure 3.4 in 1) used with conventional surgical techniques, which, in fact, mechanically eliminates any support. This is a true conceptual difference with conventional surgery: the operated forefoot must be supported (Figure 3.4 in 2).

This allows faster recovery, promotes bone healing (Figures 3.5 and 3.6), stimulates contraction of the intrinsic muscles, and improves venous circulation. These osteotomies have also changed our attitude towards central metatarsal neck fractures.

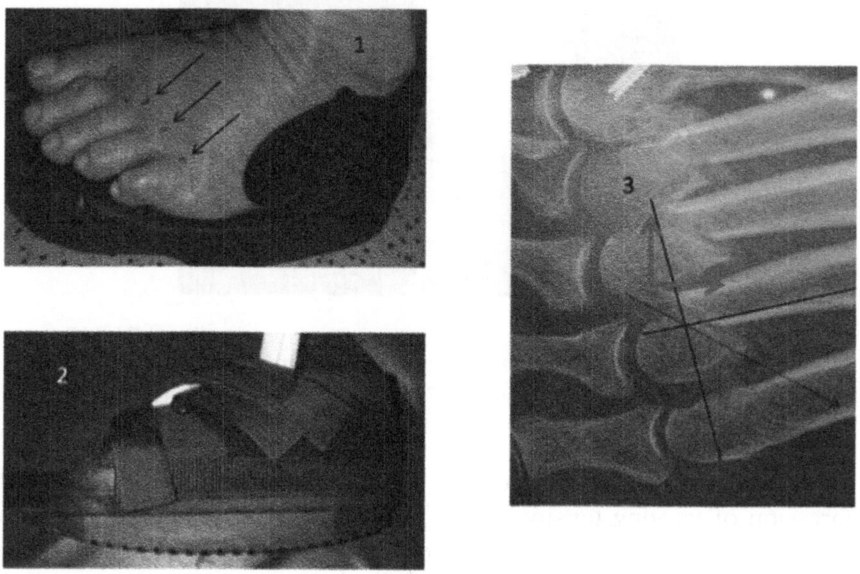

Figure 3.4. *In photo 1, the heel-supported shoe does not transfer body weight to the DMMO osteotomy area in 3, unlike the walking boot in 2*

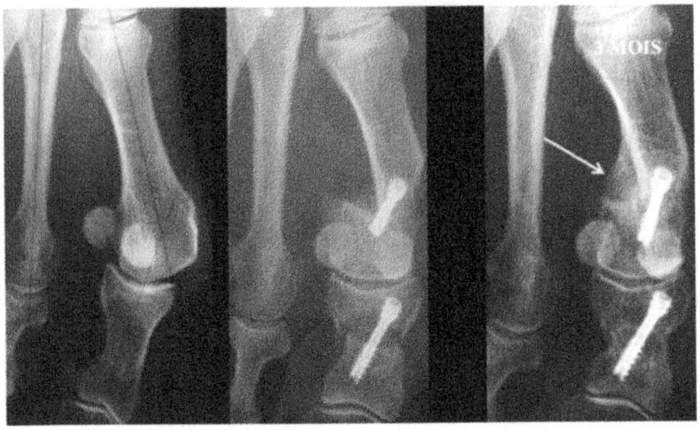

Figure 3.5. *Voluminous bony callus after a large hallux valgus that developed because the initial configuration allowing immediate total weight bearing*

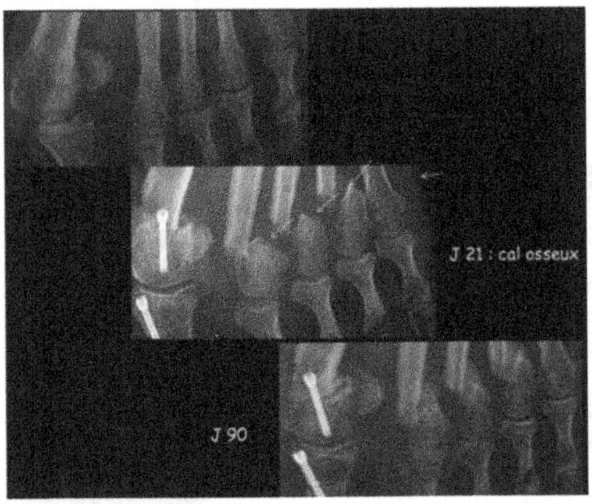

Figure 3.6. *Early and large bone callus after high amplitude DMMO for metatarsalgia in the stance phase, thanks to the initial configuration allowing immediate total weight bearing*

This posterior translation of the metatarsal heads also occurs with suppression of varying tension forces in the horizontal plane exerted on the base of the hallux, or correction of claw toes in the vertical plane. Suppression of the resultant tensile forces can also be achieved by sectioning the point of application of the tendons by tenotomies. This surgery results in dramatic corrections with minimal scarring (Figure 3.7).

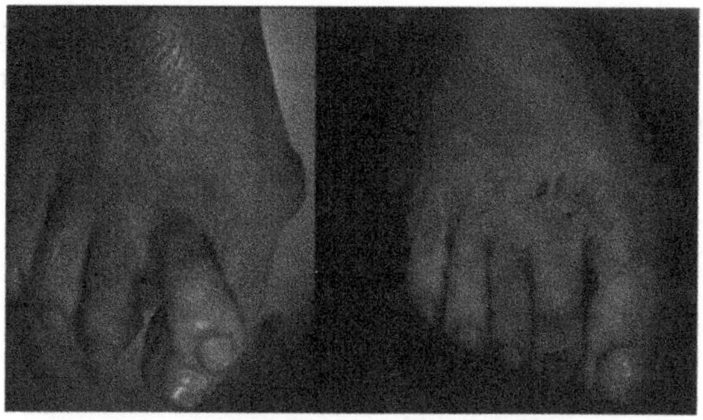

Figure 3.7. *Spectacular correction of second toe supra-adductus using the DMMO technique*

Percutaneous surgery is more than just its semantic definition. It has a heuristic function in various fields: pain management, immediate full weight bearing, manipulations, new dressing functions, technological innovation and anatomy.

The analysis of patient needs in terms of speed of recovery is the central axis around which all percutaneous choices are organized. The objective is to achieve early autonomy in daily and professional activities. The potential added value, which needs to be validated by prospective studies, would be faster functional recovery and less iatrogenesis.

3.3.4. *Absence of pain*

Pain is related to the deformation of Pacini corpuscles which are mechanoreceptors. As a very first approximation, we can estimate that pain is related to a force exerted over a certain area, in other words a pressure. Regardless of the subtleties between force, pressure and stress, this is not a biomechanics presentation.

It can be a compressive force such as lateral metatarsalgia by load transfer, by functional insufficiency of the first ray.

The second important parameter is the time over which the stresses are applied; pain is even more significant when the stress is prolonged. A patient suffering from metatarsalgia will be very sore if the duration of the digitigrade phase is prolonged by wearing high-heeled shoes.

Taking pressure and time into account (Figure 3.8) suggests that pain could be represented by a function that somehow integrates these two parameters (integral function).

$$\text{Pain: } \int_{y}^{x} P.t$$

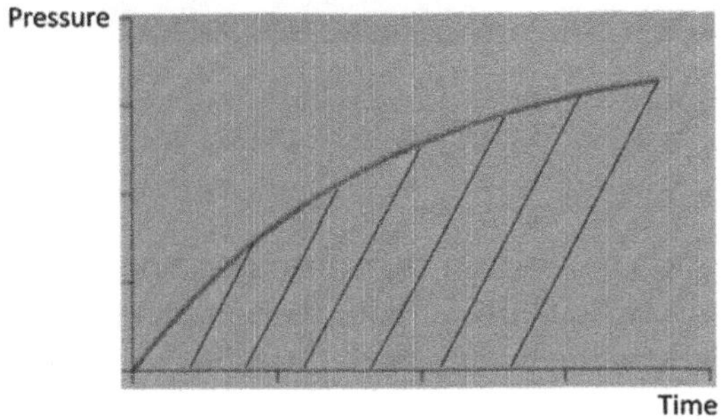

Figure 3.8. *Pain depends on pressure and application time*

The sphere of therapeutic action therefore concerns the possibilities of reducing the intensity of these mechanical stresses and reducing their application time.

The pain will also be related to the surgical technique itself, such as the mechanical soft tissue relaxation, the small size of the scars and the absence of large detachments. Outpatient surgery or short hospitalization has also been accompanied by improvements in purely sensory regional anesthesia (distal foot block) and postoperative analgesia. The foot remains anesthetized for an average of 12 hours, leaving plenty of time for effective analgesic treatment to be implemented before the foot even wakes up.

3.3.5. *Immediate full weight bearing*

Again, percutaneous surgery has allowed adjustments to be made. The generation of L.S. Barouk and P. Diebold spread the notion of osteotomy, which was a major transgression in the eyes of some of their predecessors. It was not for them to be asked to risk immediate full weight bearing. The heel-supported shoe was therefore the rule and their rule alone.

Surgeons promoting the percutaneous approach were trained at a time when the osteotomy had become mainstream and no longer represented an achievement. Under these conditions, the stable mechanical nature of the osteotomies, with or without fixation, in combination with a shoe with a rigid convex sole (Figure 3.4), simulating the gait cycle, allowed immediate complete plantar support. This confidence in the possibility of support is in fact a necessity for the required self-adjustment after *DMMO*. Not to mention the added value in terms of venous circulation, arterial microvascularization and stimulation of the intrinsic foot muscles.

3.3.6. *Manipulations*

Percutaneous surgery is inseparable from the invention of innovative manipulations. It is difficult, performed in a space that needs to be created, with limited detachments.

In contrast, arthroscopy benefits from a natural working space: the joint. Percutaneous surgery requires the simultaneous use of both hands for its execution, not only for holding the tool (in the form of a pen (Figure 3.2) and controlled by a pedal), but also for the proprioceptive information that constantly guides the surgeon. Our teachers used to say that tibial osteotomy is not a deaf surgeon's procedure, as the loud noise from osteotomes becomes metallic on the bone cortices. Percutaneous surgery is a surgery of proprioceptive perception, certainly helped by the fluoroscope, which is systematically used at the beginning of our practice and then only for delicate manipulations.

3.3.7. *Technological innovation*

Technological innovation has led to the development of tools (Figure 3.2), in particular burrs, identical in appearance to those used in ENT, that are used for trimming but above all for cutting, and replace the classic oscillating saws. The principle of sectioning by rotation is then the source of two problems: slippage during initiation and bone necrosis due the speed of rotation. The solution found was to combine decreased rotation speed with increased motor torque; these are the functions of the speed reducer and the percutaneous-specific consoles.

3.3.8. *New role for postoperative dressings*

Dressings have becoming more and more important in percutaneous surgery. More than simply providing an aseptic healing environment, it provides immobilization after osteotomy in the absence of fixation. It even allows the bone axes to be adjusted postoperatively thanks to the custom-made, modifiable silicone wedges. Finally, it contributes to analgesia and to reducing edema by keeping the toes firmly in place. It psychologically isolates the patient from the damaged area and its volume allows it to be used as a shoe at home, before the first trip outside.

3.3.9. *Outpatient surgery*

Outpatient surgery should not be an exception to the requirements of conventional hospitalization. Ensuring that the same quality criteria are met is a prerequisite for considering outpatient treatment. Nothing differs more than the technical modalities chosen by the different surgeon/anesthetist pairs.

In fact, postoperative analgesia techniques, surgical indications, the type of procedure or the incision size have so many potential combinations that it is pointless to list them all. Under these conditions, it is better to talk about what all the surgeons can agree on, a prior agreement on postoperative quality criteria, whatever the modality of management.

What are these criteria, which belong more to the realm of common sense and ethics than to medicine? The minimum required is the absence of uncontrolled postoperative pain, the absence of bleeding or extensive hematoma requiring early dressing changes and at least partial immediate autonomy being possible.

Why are there so many demands on pain management? For obvious ethical reasons, as well as because pain has the function of signaling danger and will be interpreted as a signal of fragility, discouraging the patient from using the operated organ. Hematoma or extensive postoperative bleeding on return home must be avoided. In the most serious cases, they will cause an immediate return to the operating room. In less serious cases, they will fuel the patient's anxiety and lead to the same result as the presence of pain: delay in using the operated organ and thereby in functional recovery. Concerning immediate autonomy, it is as important to the patient for daily life as it is in case of emergency, to avoid any danger.

Outpatient surgery should only be performed if it is certain that these criteria are met, *a minima*.

How can we be sure that the post-operational effects will fulfill the above-mentioned criteria? Will this certainty be acquired in the choice of pathology to be treated, by the patient himself or by the technique used?

How do we prioritize the cited selection criteria?

Contraindications are the starting point for a sorting process that makes it possible to comply with the prior agreement defined in the introduction. The aim of this screening is to obtain what seems to be the minimum objective that all surgeons can agree on.

Mandatory postoperative criteria must be met in outpatient treatment. In order to determine the cases likely to benefit from them, it seems simpler to start with the contraindications that fall within the three categories: background, pathology and technique.

The less aggressive the procedures carried out, the more it seems possible to treat severe pathologies. The choice of technique partly determines whether outpatient treatment is feasible. We must take into account the type of operation, the patient's aptitude for hospitalization on an outpatient basis, but also the surgeon's capacity. These elements are more decisive in the choice of outpatient treatment, with the pathology to be treated taking second place.

The choice of outpatient surgery is not the subject of a dogma that necessarily attributes a pathology to this type of treatment. Rather, the dogma lies in the postoperative achievement of qualitative criteria related to pain, bleeding and autonomy.

Under these conditions, a preliminary triage must be carried out, the nature of which depends on surgical and anesthetic skills, as well as and above all on the background. For economic reasons, this triage must not depend in any way on external forces such as paying agencies or health care institutions. Patients themselves are sometimes very demanding for practical reasons, but also because the idea of going out the same day confers a dimension of harmlessness. In these circumstances, they need to be reassured without endangering them.

3.3.10. *New work in anatomy*

The fate of anatomists and surgeons had diverged for a few generations. Here, percutaneous surgery has made it possible to formulate a request for an original anatomical study, difficult to carry out, consisting of defining micro approaches, avoiding the dangers of injury to critical structures. The remarkable scientific and artistic work of Pau Golano, an anatomist in Barcelona, who died suddenly in 2014, is to be commended. This work forms the conceptual and gestural basis of percutaneous surgery.

3.3.11. *Conclusion*

To a certain extent, percutaneous and minimally invasive foot surgery must be considered as an efficient technical evolution of conventional surgery. The second aspect is more original and constitutes a new concept.

Percutaneous surgery has unquestionably revitalized surgical practices with a minimally invasive approach, whether in the choice of approaches, the invention of original manipulations, the possibility of real weight bearing, pain management, anatomical research and individualized geometric adjustment.

It is in this respect that the term *percutaneous* should be considered as an ellipsis to a much broader representation system.

3.4. Biomechanics of stability

To achieve fast healing the surgically altered organ must benefit from immediate mechanical loading, the source of tissue growth. The most intelligent way to proceed is to break down the loads by classifying them as dangerous or harmless. In the case at hand, that of postoperative foot osteotomies involving small bones, the compressive stress induced by immediate full weight bearing stimulates the bone (Figures 3.5 and 3.6) and presents no danger. On the other hand, flexion moment are prohibited for three weeks, neutralized by a postoperative walking boot (Figure 3.4), which does not allow the toes to bend, and simulates the gait cycle with full support.

We will illustrate how to obtain solidness that allows immediate postoperative support even before biological bone healing, by presenting a technique for correcting hallux valgus called a chevron osteotomy. This V-shaped osteotomy is done at the distal part of the first metatarsal, resulting in the bony parts being embedded together (Figures 2.9b and 3.9).

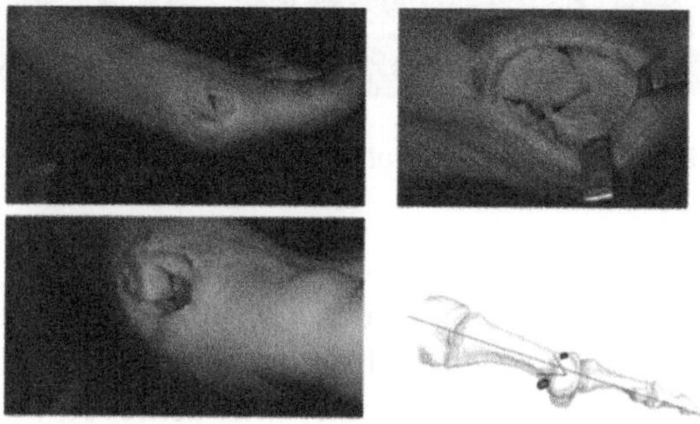

Figure 3.9. *V-shaped chevron osteotomy with stable embedding. For a color version of the figure, see www.iste.co.uk/cazeau/foot.zip*

Furthermore, the osteotomy is performed in strong cancellous bone, as it is located in the metaphyseal zone. The cuts are located close to the center of rotation, thus minimizing the flexion moments (Figure 3.10).

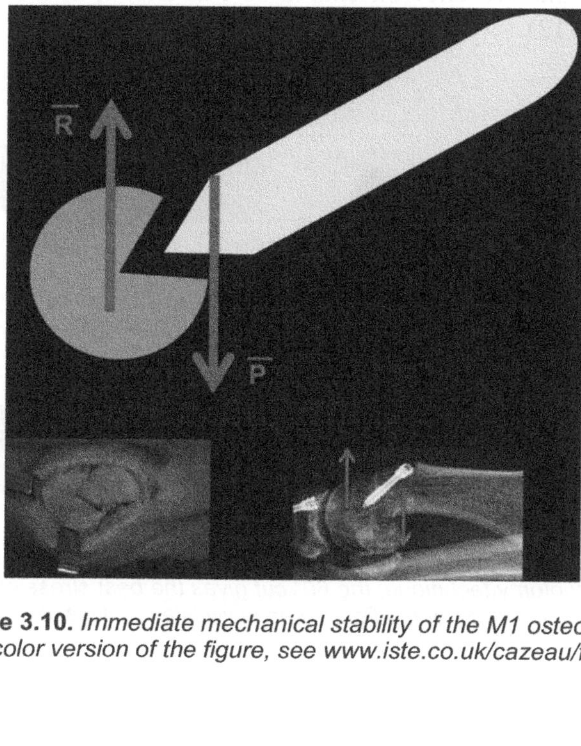

Figure 3.10. *Immediate mechanical stability of the M1 osteotomy. For a color version of the figure, see www.iste.co.uk/cazeau/foot.zip*

This is supplement by fixation, represented here by a screw with a double start thread that compresses the assembly (Figure 3.11).

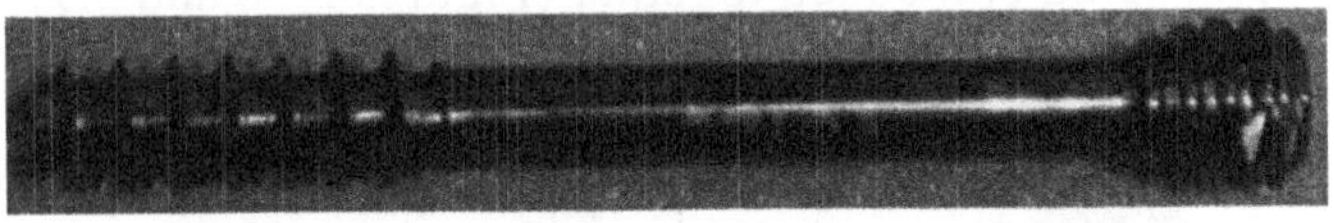

Figure 3.11. *Titanium bone fixation screw, with differential threads (tighter distally) to compress the osteotomies, thus increasing stability*

Figure 3.5 shows a chevron procedure for hallux valgus; early and satisfactory bone callus is noted with immediate support. Despite the large anatomical correction, the initial stability was achieved by the geometry of the bone cuts.

Figure 3.6 shows a *DMMO* procedure with large metatarsal head displacement; the initial stability allows support and early appearance of bone calluses.

The peak stresses in the bone are low and better distributed when the angle between the two chevron cuts is 60°, as shown in the finite element study (Figure 3.12).

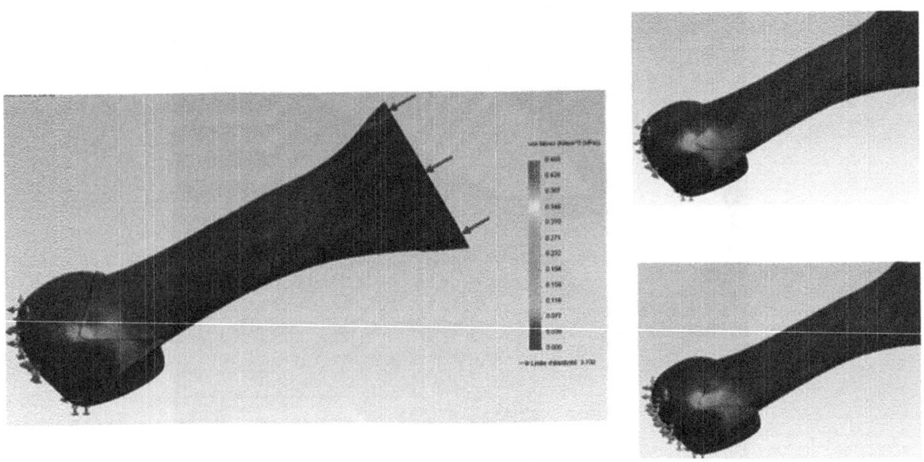

Figure 3.12. *Computer finite element modeling of different cutting angles in the M1 chevron osteotomy technique; the 60° cut gives the best stress distribution (left). For a color version of the figure, see www.iste.co.uk/cazeau/foot.zip*

Conversely, a vertical and flat section like a sausage (Figure 3.13) is very fragile and does not provide good support, due to shear forces placed in the frontal plane.

For the first phalanx, the realignment technique producing immediate stability is related to the preservation of a lateral bone wall serving as a hinge. Strength is further enhanced by inserting a screw on the opposite side (Figure 3.14).

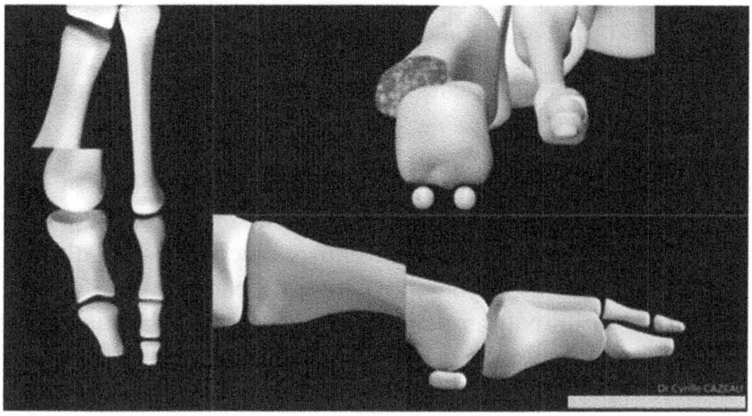

Figure 3.13. *Bone cut made in the frontal plane, which is unstable. For a color version of the figure, see www.iste.co.uk/cazeau/foot.zip*

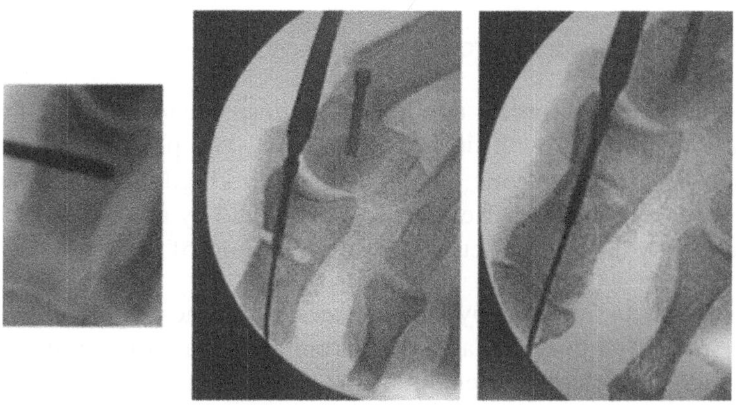

Figure 3.14. *P1 varus osteotomy for hallux valgus, using a motorized burr*

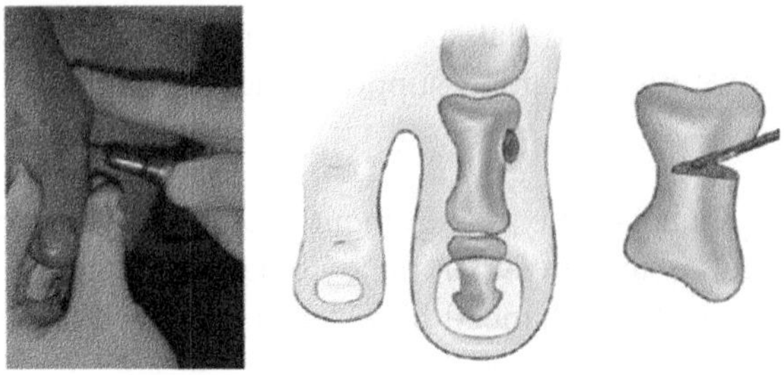

Figure 3.15. *P1 varus osteotomy for hallux valgus; the technique is percutaneous via a 2 mm skin incision*

3.5. Pain and recovery

Any postoperative pain is interpreted as a sign of fragility by the patient. Indeed, pain is an evolutionary survival mechanism that triggers the organ to be removed from the nociceptive agent. Since mechanostimulation is a major factor in bone growth, the removal of weight bearing slows down healing. Measures such as minimal surgical damage and initial mechanical stability are then useless. The absence of pain is inseparable from the two previous ones. To solve this problem, anesthesia should be long enough to allow effective oral treatment to be implemented. In order to allow walking with full weight bearing, this anesthesia must not inhibit motor activity. A purely sensory local block of the foot only is needed that is prolonged.

3.6. Forces and tissue growth

The mechanical forces applied to living organisms on this earth contribute to tissue growth, in both the animal and plant kingdoms. These forces consist of weight due to gravity in both kingdoms and other forces specific to each kingdom. These are forces applied by muscles on the animal skeleton and by the wind's action for aerial portions of plants.

Tissue growth induced by these forces intervenes directly on survival mechanisms. In animals, it allows appropriate repair of the skeleton, and plays a part in posture, orientation and movement. In plants, it helps to reinforce the areas that are mechanically weak and helps with growth, posture control, resistance to forces related to its own weight and the lift of its foliage in the wind.

The mechanisms that translate these forces into tissue growth are not fully understood. Nevertheless, there is strong analogy between the animal and plant kingdoms, and the overall functional scheme is the same. The initial step is the stimulation of mechanoreceptors. The forces are then transmitted to cells that bathe in extracellular matrix by the intermediary of a rigid intracellular conduction system – the cytoskeleton. A complex biochemical reaction ensues, resulting in tissue growth that meets the needs of the organism, whether animal or plant. The resulting quantitative and qualitative macroscopic changes in the supporting elements – bone for animals and wood for plants – will alter their physical properties to match survival requirements.

3.6.1. *Introduction*

About 400 million years ago, the first forms of life on Earth were plants. Originally aquatic, they colonized continents and subsequently were subjected to mechanical forces much stronger than in water. They adapted in such a way that roots grew into the soil and aerial elements grew towards the sky. Under these conditions, wood must be sufficiently rigid to prevent failure while also being flexible enough to absorb plastic and elastic energy.

Phylogeny and ontogeny had to make them "right", therefore adapt to the environmental mechanical constraints (Seeman 2003). These stresses are a key signal affecting morphogenesis and growth. A distinction is made between endogenous forces such as muscle forces in animals, elastic tension or turgidity in plant stems and external forces. The latter consist essentially of gravity in both kingdoms, touch (thigmotaxis) and wind in plants. Present from the origins of life, gravity is unavoidable, constant, with fixed direction that is both an unchanging reference and a physical restriction factor (Anken 2003). Present before life, it has accompanied the whole phylogenesis. The question raised here is how this vector influences ontogenesis during growth and development. However, plant growth is possible without wind (roots) and without gravity (space).

How do plants adapt to a mechanical change in their environment given that they cannot escape it (Moulia *et al.* 2006)? Plants are subjected to various mechanical forces through their relative immobility, in an aerial environment with low buoyancy and the forces of gravity and wind lift. They must fight against their own weight, the instability linked to their growth upwards to concentrate solar energy through photosynthesis, and downwards through the root system to capture water, nitrogen and mineral salts essential to their growth. In a tropical forest, a tree in a dark undergrowth must grow rapidly to reach the light at the top of the canopy, and then it must resist the wind forces acting on the foliage it develops. At the same time, it must retain its root geotropism.

Conversely, animals can flee environmental constraints, but to do so they must have a rigid and articulated skeleton, and be able to fight against gravity by maintaining itself in an erect position through muscular forces, which also contributes to its mobility.

Francis Hallé (2014) introduced an original notion – the relationship between the shape of living beings and the space in which they live. The space classified as three dimensional according to Euclid must be resized by its verticality determined by the omnipresent factor of gravity. Gravity introduces a vertical polarity (Figure 3.16).

Figure 3.16. *Vertical growth of a larch resumes despite the slope*

That of plants goes from the root system to the crown, according to radial symmetry, with polarity and symmetry being carried by the same vertical axis. Animals have two polarities: one is linked to gravity, materialized by dorsiventrality, and the other is apicocaudal, linked to displacements, as well as bilateral symmetry. The shape of multicellular living beings is strongly conditioned by gravity (Thompson *et al.* 2009), but not exclusively; the shape of plants also depends on the search for light and the morphology of animals in search of mobility. When ecological factors such as humidity, heat and sunshine are abundant, as they are in tropical rainforests, they do not constitute limiting elements; only the heavy physical constraint of gravity is present, allowing the observation of all known structures respecting symmetries and vertical polarities (Hallé 2014). Conversely, living beings that are not subject to gravity are those whose mass is infinitesimal (foraminifera, etc.). Subject to the forces of surface tension, viscosity, Brownian motion and electrical charges, the above descriptions do not apply to them because of their symmetries and polarities (Thompson *et al.* 2009).

Here, we will try to draw analogies between the plant and animal kingdoms in terms of gravity's mechanisms of action. "The search for relationships between apparently independent things and the search for similarities between dissimilar things in the eyes of mortals" is a classic theme of Aristotelian thought.

3.6.2. *Applicable forces*

3.6.2.1. *Gravity*

3.6.2.1.1. Lifespan

Gravity was present before the onset of life and probably influenced the entire evolution (Volkmann and Baluška 2006; Najrana and Sanchez-Esteban 2016; Turko *et al.* 2017). The load on land is a thousand times greater than in water, according to Archimede's principle, and ecological factors are much more varied (temperature, distribution of nutrients, radiation). The mechanical anti-gravitational response on land is thus much more significant, whether through animal ossification or plant turgescence and lignification. Experiments on amphibian fish show that their terrestrial acclimatization is accompanied by skeletal reinforcement reflecting an increase in weight-related strength. This suggests that weight detection is an ancestral trait, not an innovation that accompanied the development of terrestrial tetrapods. In addition, current amphibians such as frogs cannot remain in the aquatic environment (microgravity) all the time, as they will lose bone mass due to osteoclastic degradation (Besova *et al.* 1993).

Gravity is perceptible by today's most sophisticated organisms (Najrana and Sanchez-Esteban 2016). This also argues in favor of the lifespan of its influence; in space, bacterial growth increases by 90%, and the expression of prokaryotic genes or the growth of yeasts are modified. Gravity is also expressed in some very old zoological families such as jellyfish (Singla 1983), sponges and the close ctenophores. These generally possess a large and unique statolith, which is a solid body involved in the perception of gravity. It is located in a cell, the statocyst, and thus controls the locomotor system to maintain the animal in the direction of gravity (Tamm 2014). This system is the equivalent of our otoliths positioned in the vestibular apparatus. Another "living fossil", the nautilus, is a marine cephalopod (like the octopus, squid, cuttlefish) that has not evolved for 400 million years, and has statocysts that allow it to adapt its orientation and movements to gravity, sensitive to angular accelerations (Neumeister and Budelmann 1997).

3.6.2.1.2. Intricacy between rigid bone skeleton and calcium homeostasis

Calcemia is tightly regulated in the organism because it plays a critical role in membrane excitability and coagulation. The ability to maintain homeostasis arose at the same time as life developed, and its mechanisms changed during the transition from water to land due to the lack of water and ions available in non-aquatic environments. In parallel and in interaction, bone skeletons capable of withstanding the forces of gravity in the terrestrial environment were built up. Bone contains about 1 kg of calcium in the form of hydroxyapatite crystals in humans (25,000 mmol). This reserve is vital to maintaining constant blood calcium levels in the event of a deficiency and is used as a skeleton.

Here is another example that illustrates both the intertwining and the life span of these two phenomena. Tunicates are marine invertebrates with a similar structure to the chord of vertebrates (embryonic tissue, at the origin of vertebrae) and have evolved very rapidly since the Cambrian period. They have the calcium-sensing receptor (CaSR) present in humans, which is used for the feedback control of parathyroid hormone in calcium homeostasis.

Already present in evolution, it is inactive since there is no need for calcium control in a stable marine environment of water and ions.

The passage on earth, and being subjected to the forces of gravity, have linked the control of calcium homeostasis to the sophistication of skeletal bone remodeling.

3.6.2.1.3. Organism-wide dissemination

Plant tissues

Figure 3.17. *Resumption of downward vertical growth of a root tip that was horizontalized experimentally (Cazeau 2018, Maitrise Orthopédique)*

In plants, the best known mechanoreceptors are the statolites located in part of the roots, the columella, which are present in other structures of the plant stem. They allow reorientation of the roots towards the soil (Figure 3.17), growth of the wood, its repair, increase in strength under the effect of external forces, vertical orientation of the young shoots during reiteration, etc. (Figure 3.18a and b).

A key structure, the meristem (Figure 3.19) is a structured multicellular zone responsible for the growth of all plant organs.

Contrary to the beliefs in Antiquity and the Renaissance, each living being is not entirely pre-formed from the embryo (theory of pre-formation), but is built up through conflicts with the external environment (theory of epigenesis, Aristotle, Goethe and Caspar Wolff). The meristem is particularly sensitive to mechanical stresses, constituting a signal for the cells forming the organs, via an auxin gradient. Others are responsible for a certain mobility, such as trapping carnivorous plants.

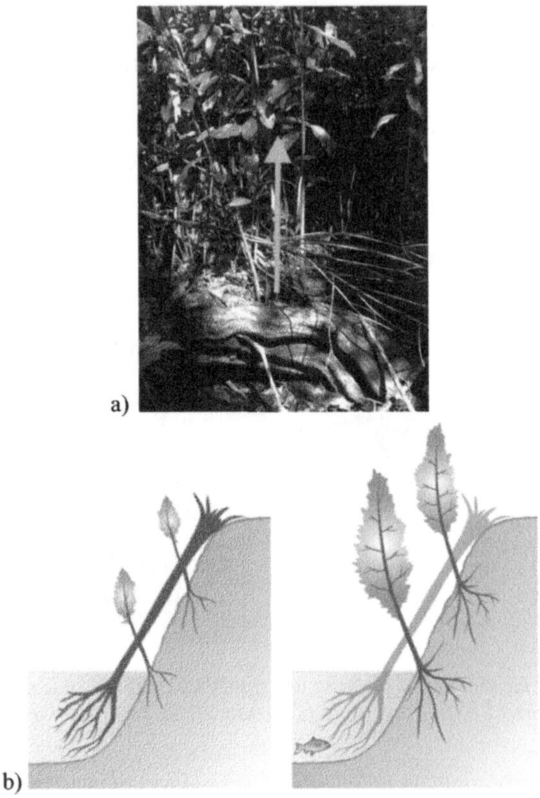

Figure 3.18. *a) Resumption of vertical upward growth of young shoots on a stump felled in the Guyana forest (reiteration). b) Redrawn from Éloge des plantes by Francis Hallé*

Figure 3.19. *The meristem – the plant's growth zone – is at the origin of all its organs*

Bone tissues

In animals, the effects of gravity are identified by numerous observations, particularly in situations where the force of gravity is absent. The action on the mechanoreceptors of osteocytes results in biochemical reactions that cause growth with bone calcium deposits.

Conversely, bone mass decreases by up to 2% per month in space due to lack of gravity, although resistance exercise reduces these losses (Holick 1998).

Other studies show that a 4- to 6-month stay in space causes astronauts to lose 1–14% of their bone mass depending on the location, mainly load-bearing bones: spine, pelvis, femoral neck and trochanters (Heer *et al.* 1999). Fractures heal more slowly and the mechanical properties of the bones decrease (Gadomski *et al.* 2014). Partial loading such as lunar gravity (one-third of that of the Earth) reduces bone and muscle loss (Swift *et al.* 2013). PTH supplementation is ineffective (Fukumoto 2008; Sakai 2012).

Conversely, hypergravity has deleterious effects above 3 G (Gnyubkin *et al.* 2015).

Heart, circulatory system and lungs

In microgravity, fluids move to the upper part of the body, affecting cardiac filling, the ventilation/perfusion ratio (Prisk 2014) and oxidative stress (Takahashi *et al.* 2017).

One explanation is the decrease in fluid pressure in the interstitial compartment, which may be involved in maintaining bone mass. It seems that osteocytes do not react directly to deformation under load but to changes in extracellular fluid pressures. This is suggested during prolonged bed rest (Lanyon 2009), where bone loss is observed especially near the feet, while cranial bone mass increases (Leblanc *et al.* 1990).

Increased venous pressure leads to increased periosteal bone formation in dogs, possibly by capillary hyperinfusion of osteoblast precursor cells (Kelly and Bronk 1990).

In space, capillary vessels atrophy, bone stem cells transform into fibroblasts and adipocytes, leading to fibrosis without mineralization

(Rodionova and Oganov 2003). Red blood cell mass has been found to decrease by 10–15% after a stay in space (Talbot and Fischer 1986).

Muscle tissues

In the absence of gravity, muscle mass decreases, but this does not explain the intensity of bone demineralization. Muscles seem to act more by maintaining an erect position that allows them to be subjected to gravity than by their direct action on bone (Laurent *et al.* 2016). This is confirmed by comparing the effects of muscle inhibition by botulinum toxin, with or without gravity (Warden *et al.* 2013). Russian experiments during prolonged stays in space showed that despite implementing physical exercise protocols, there was a significant increase in urinary calcium loss, especially in the spine, accompanied by muscle loss in the lower limbs (Whalen 1993). Physical exercise in space maintains bone forming activity but resorption is increased (Vico *et al.* 1987). In the fetus in microgravity, there is a decrease in muscle mass in the lower limbs, as well as a decrease in the bone mass of the bones used to achieve the erect position, and a decrease in cardiac function (Sekulić *et al.* 2005).

Nervous system and vestibule

The effects are complex. Changes in gravitational forces in the vestibule are detected by otoliths (Figure 3.20) in the hair cells of the inner ear (Najrana and Sanchez-Esteban 2016). This allows the control of posture, body and spatial orientation.

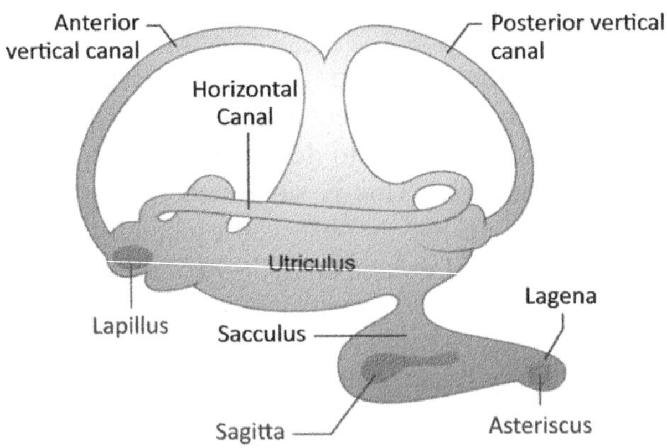

Figure 3.20. *Vestibule of the inner ear of a fish with three statoliths (Cazeau 2018, Maitrise Orthopédique). For a color version of the figure, see www.iste.co.uk/cazeau/foot.zip*

The information is transmitted to the suprachiasmatic nuclei with consequences on the nycthemeral rhythm, diminishing the organism's ability to adapt to any form of stress. Conversely, the brain plays an important role in the mineralization of fish otoliths and depends on gravity (Anken 2006). The sympathetic system's activity is altered (Kawao *et al.* 2016).

Genetics

Gene expression at the cellular level is affected, with a notable influence on the boreal rhythm and all its physiological consequences (Czeisler *et al.* 1991; Whitson *et al.* 1995; Fuller *et al.* 2000; Murakami and Fuller 2000; Robinson and Fuller 2000).

Stem cells

The fate of mesenchymal stem cells is altered (Najrana and Sanchez-Esteban 2016) and does not result in osteogenesis (Shi *et al.* 2010). Bone marrow cells remain undifferentiated in weightlessness and could therefore still be transplanted (Yuge *et al.* 2011).

3.6.2.1.4. Diffusion throughout animal kingdom

Mechanoreceptors are present in all taxonomic ranks of the animal kingdom. They are found in all vertebrates, mammals, fish, amphibians, reptiles and birds. It is interesting to note that they are absent only in fish with a non-bony but cartilaginous skeleton (chondrichthyes) like sharks.

They are also present in invertebrates, such as mollusks, insects and crustaceans (arthropods), annelids, cephalopods, etc.

These include work in the literature on cuttlefish, octopus, nautilus and squid (Solé *et al.* 2013). Within arthropods, the presence of statocysts has been studied in hymenopteran insects (Ishay *et al.* 1983) and many marine (Hama *et al.* 2007) and freshwater crustaceans (Schöne and Steinbrecht 1968), evidence of their role in posture control.

In growing amphibian vertebrates, the statocyst has been studied in weightlessness in space, showing similar function to invertebrates (Vinnikov *et al.* 1974). Studies have also been carried out on numerous marine gastropods (Pedrozo *et al.* 1996; Wiederhold *et al.* 2000), in space and on land, showing the influence of gravity on the control of vertical posture, and locomotion by contraction of the "foot". Some papers show that the statolith mass increases in space (microgravity) and decreases in hypergravity, as if the force of weight formation were to remain the same (Pedrozo *et al.* 1996). Other studies also focus on the terrestrial snail by showing how equilibrium changes in weightlessness via statocysts (Aseyev *et al.* 2017). Statoliths are also present in annelids and helminths. The sandworm (Kharkeevitch and Titova 1974) is able, through two statocysts, to recognize its position in space 40 cm below ground.

3.6.2.2. *Associated forces*

3.6.2.2.1. Animals

Muscular forces applied to bone have a direct effect on bone remodeling but much less significant than the effects of gravity. However, they have an indirect role because they make upright posture possible through their traction such as rope ladders or cables, and thus allow the force of weight to be expressed through the reaction on the ground. They therefore make the force of gravity effective (Thompson *et al.* 2009).

3.6.2.2.2. Plants

These are the carrying forces transmitted to the branches by the action of the wind (Figure 3.21).

Figure 3.21. *Deformation of branches under wind forces; "flag tree" (Cyrille Cazeau)*

They can have spectacular effects through the formation of a reaction wood adapted to withstand the stresses. Excess wind forces can lead to fracture or uprooting (Figure 3.22a and b).

a)

b)

Figure 3.22. *A strong wind can cause a) uprooting of the tree or b) fracturing of its base (Cyrille Cazeau)*

Their absence during growth makes the tree vulnerable. Experimental cultivation of a tree sheltered from the wind makes the trunk straight, narrow, fragile and fast-growing. Installation of a horticultural stake makes the trunk tall and thin. These forces are significant; in wind-beaten areas, the trunks are short and stocky. The stem detects bending of its base due to the wind and modifies its growth. There is an increase in root diameter.

The consequences of touch on growth (thigmomorphogenesis) are not described in this chapter. It takes too long to describe the nuances with thigmotactism and nastic movement.

3.6.3. *Mechanoreceptors – the starting point*

3.6.3.1. *Animals*

3.6.3.1.1. Otocyst and otoliths

The vestibular system in zoology is well known. Otoliths (Figure 3.20) present in the utricle and sacculus of the vestibule are sensitive to gravitational forces, with movement, allowing detection of accelerations and gravity (Najrana and Sanchez-Esteban 2016). They excite sensory cells whose cilia are formed of actin filaments. Next, a biochemical cascade is activated, sending information to the brain. Several experiments in space show that the mass of the otoliths changes in order to keep the gravitational information identical, and thus constant body dimensions (Walton *et al.* 1992; Pedrozo *et al.* 1996; Moody and Golden 2000; Ronca *et al.* 2000; Wubbels *et al.* 2002; Ronca *et al.* 2008). The site of calcium incorporation is located directly adjacent to the sensory epithelium and is under the influence of gravity through the brain. Cutting the vestibular nerves will stop calcification of the otoliths (Edelmann *et al.* 2004; Sekulić *et al.* 2005).

3.6.3.1.2. Osteocyte

There are three main cell types in bone (Iolascon *et al.* 2013): osteoblast, osteoclast and osteocyte (Figure 3.23).

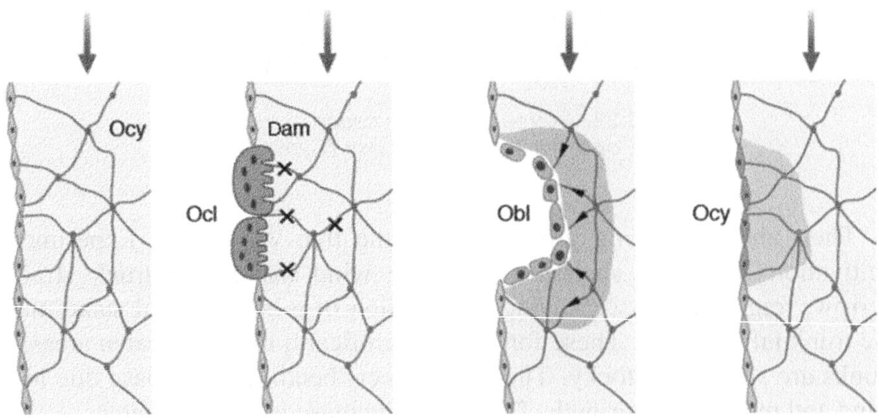

Figure 3.23. *The three types of bone cells: Ocy: osteocytes; Ocl: osteoclasts; Obl: osteoblasts; Dam: damage, microlesions due to mechanical stress induced by a vertical force (Cazeau 2018, Maitrise Orthopédique). For a color version of the figure, see www.iste.co.uk/cazeau/foot.zip*

The first two will allow bone remodeling for permanent maintenance by replacing an old bone with a young bone. Osteoclasts, which degrade bones, are placed into action, then the osteoblasts (collagen) produce an extracellular matrix, which is mineralized by numerous hormones. Cellular interactions take place via several mediators (RANK system, osteoprotegerin, etc.). The osteoblast dies in a programmed manner (apoptosis), allowing itself to be trapped in the bone matrix that it developed; it then transforms itself into a third player, the osteocyte.

Its cell body is located in a bone gap surrounded by extracellular matrix, and is extended by dendrites located in the canaliculi (Figure 3.24).

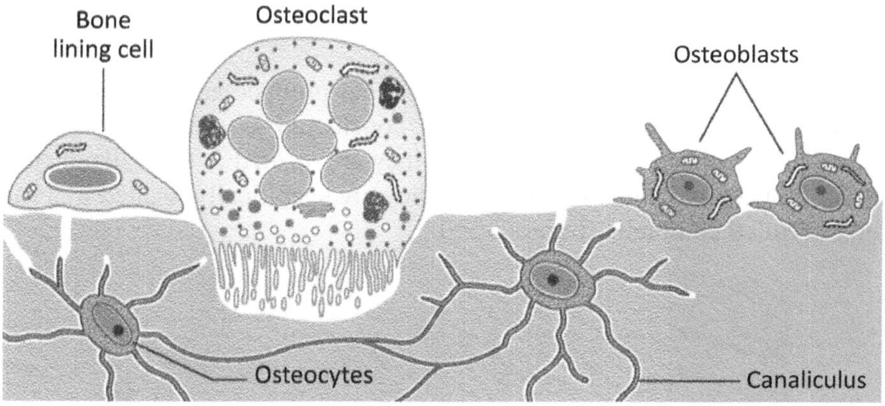

Figure 3.24. *Osteocytes communicate through their dendritic connections in the lacuno-canalicular network (Cazeau 2018, Maitrise Orthopédique). For a color version of the figure, see www.iste.co.uk/cazeau/foot.zip*

Dendritic processes are linked by connexins – joining proteins – which are mechanosensitive. The dendritic network in this lacuno-canalicular network allows communication between osteocytes, osteoblasts and other cells. The cell surface is ciliated, sensitive to extracellular fluid movement, thus intervening in the flow of calcium, as in the inner ear. Their main role is to detect surface microcracks and initiate the remodeling process. The stimuli are the presence of stress/strain in shear and stretching, as well as lacuno-canalicular pressure modifications. It appears that osteocytes do not react directly to deformation under load but to changes in extracellular fluid pressure as suggested by cardiopulmonary changes that occur in weightlessness (Burger and Klein-Nulend 1999). They play a central role in detecting mechanical signals and regulating bone formation and resorption through osteoblasts and osteoclasts (Evans 1976; Najrana and Sanchez-Esteban 2016).

In microgravity, the reduction of deformities causes bone loss (Milgrom *et al.* 1985). Mesenchymal stem cells differentiate into osteogenic cells when subjected to mechanical stimulation, but induce adipogenesis in its absence (Khan *et al.* 1995), and are directed to produce the osteoblasts responsible for bone loss (Burr *et al.* 1985), and to increase the number of osteoclasts (Rubin *et al.* 2000).

3.6.3.2. *Plants*

Amyloplasts are intracellular structures that produce and store starch by the polymerization of glucose: they are considered as statoliths. They can sediment in the direction of gravity when they are located in specialized cells, the statocysts, comparable to otocysts. They are specifically located in a root structure called columella (Figure 3.25).

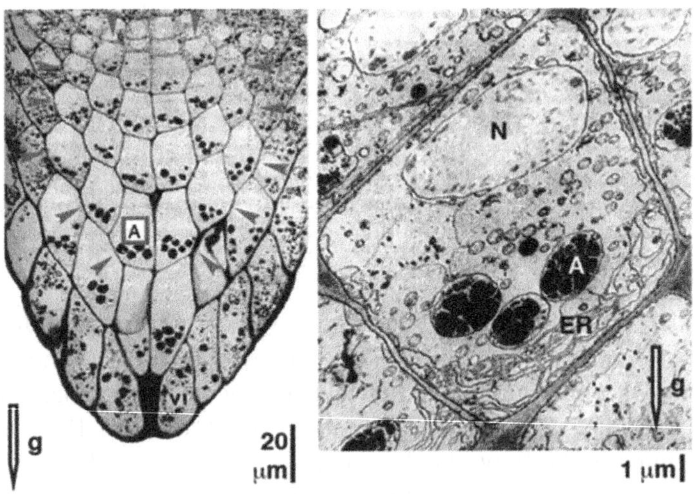

Figure 3.25. *The root tip contains an area called columella with cells containing the amyloplasts (A) that function as vestibular otoliths. For a color version of the figure, see www.iste.co.uk/cazeau/foot.zip*

The relationship between amyloplast displacement and response to gravity has been shown in several studies (Sack and Leopold 1985; Kuznetsov 1997; Fitzelle and Kiss 2001).

If root orientation is artificially altered, the amyloplasts sediment in another basal region, inducing differential root-to-soil growth (Figure 3.17). Displacement of statoliths *in vitro* using laser tweezers against the endoplasmic reticulum results in altered calcium fluxes (Leitz *et al.* 2009). A biochemical cascade, producing auxin, among other things, allows the formation of curves reorienting the roots in the soil according to gravity by asymmetric growth. This phenomenon affects other structures in the flower stem. The analogy with animals is obvious.

3.6.4. *Intracellular transmission: cytoskeleton and extracellular matrix*

3.6.4.1. *Animals*

Here we speak of animals with bony skeletons.

3.6.4.1.1. Cytoskeleton

Osteocytes are connected to each other and to surface cells by a dendritic network (Figure 3.23). They can extend and retract these cells, and thus cause cellular deformations by acting on the cytoskeleton. The cytoskeleton is composed of actin (tensile strength), microtubules (very high compressive strength) and intermediate filaments that play a role in stability (Figure 3.26).

The mechanical force at the mechanoreceptors is transmitted to the cytoskeleton which deforms and then results in biochemical reactions. For example, there is evidence that the position and shape of microtubules and filaments are altered as a function of gravitational forces (Papaseit *et al.* 2000; Tabony *et al.* 2007). Conversely, microgravity destroys the architecture of the cytoskeleton by suppressing the osteocytes' dendritic junctions (Di *et al.* 2011).

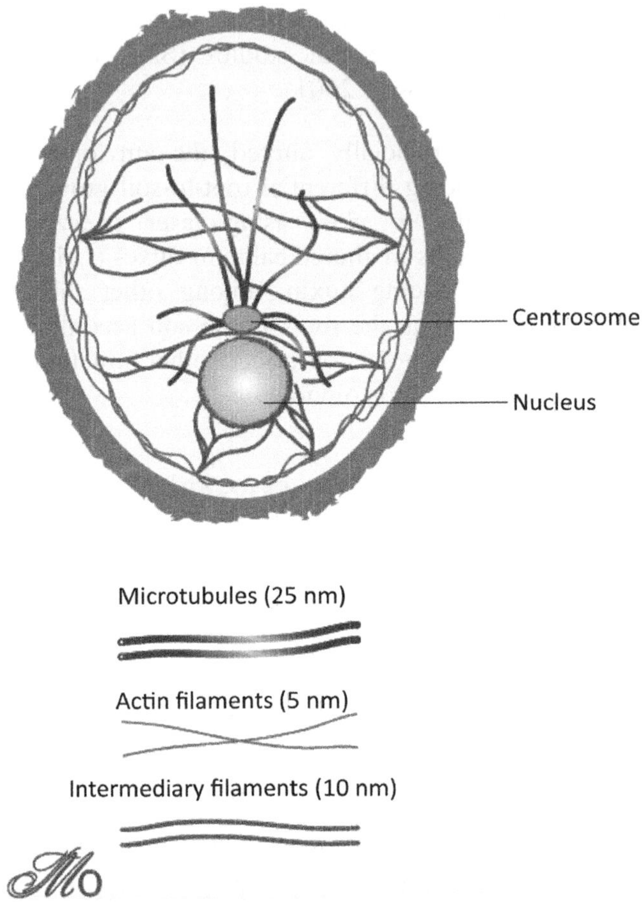

Figure 3.26. *The cell contains three types of fibers that make up its cytoskeleton, whose role is to relay external forces within the cell (source: Cazeau 2018, Maitrise Orthopédique). For a color version of the figure, see www.iste.co.uk/cazeau/foot.zip*

This has been studied experimentally for the reorganization of actin fibers (You *et al.* 2001; Han *et al.* 2004).

The cytoskeleton can be compared to a model known in the architectural world as "tensegrity" The role of the cytoskeleton can be explained by the model known in architecture as tensegrity (Ingber 1998): a three-dimensional structure consisting of tensional and compressional elements configured to resemble a trellis in space describes a self-stressed system consisting of assembled elements, giving the appearance of a trellis in space (Figure 3.27a and b).

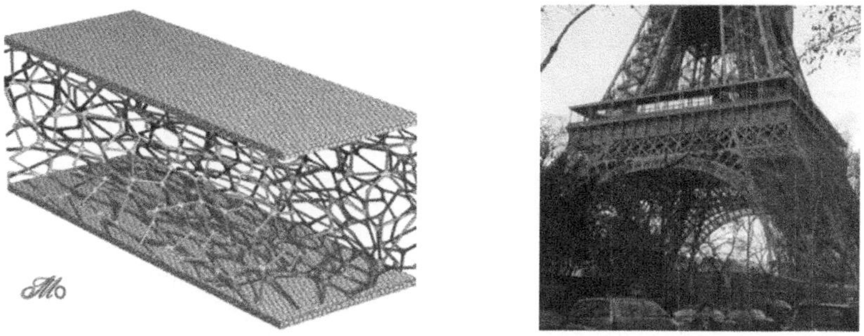

a) b)

Figure 3.27. *a) Trellis model illustrating the tensegrity structure concept, which explains the distribution of forces and self-stability. b) Architecture of the Eiffel Tower (Cyrille Cazeau)*

It is the sum of forces applied to the cellular skeleton that solidly holds the shape of each element, and then of the entire unit. Thus, it is likely that all cells are sensitive to changes in gravity, by the intermediary of this tensegrity structure, not by the direct activation of a single receptor.

3.6.4.1.2. Extracellular matrix

The extracellular matrix is mainly made of collagen fibers that resist traction, and other components such as hydroxyapatite, a substrate of bone mineralization. Based on *in vitro* studies, it appears that osteocytes react more to shear forces in the extracellular interstitial fluid than to direct mechanical deformation, and that the extracellular matrix acts as an amplifier of the mechanical signals transmitted to the cytoskeleton. The proteins of the pericellular matrix, in particular glycocalyx, are involved in changing the organization of the cytoskeleton's actin, as transducers of mechanical signals (Weinbaum *et al.* 2003).

3.6.4.2. *Plants*

3.6.4.2.1. Cytoskeleton

Cellular tissues have high mechanical properties, such as turgescent (Figure 3.28), hydrostatic or wood tissues.

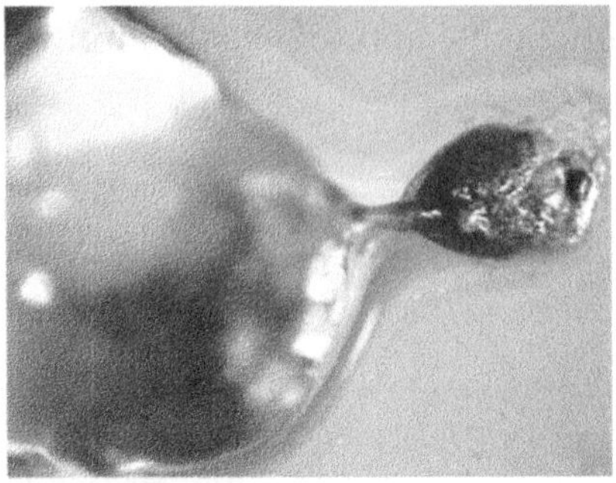

Figure 3.28. *Animal hydroskeleton consisting of a turgescent tissue, here a retractable slug tentacle with an eye at its tip (Cyrille Cazeau). For a color version of the figure, see www.iste.co.uk/cazeau/foot.zip*

Since the shape of the cell is mainly defined by the osmotic pressure/wall resistance interaction, the cytoskeleton, which is nonetheless present, is of lesser importance. There are no intermediate filaments, but the microtubules network is highly developed and is organized into a structure that presses against the membrane in conjunction with the cellulose fibers of the wall.

Among the resulting biochemical reactions, certain proteins (ARG1, RHG) are associated with the actin skeleton. These filaments probably act as relays for the intracellular transmission of forces transmitted by the mechanoreceptors. These interactions have been demonstrated in space (Volkmann *et al.* 1999; Driss-Ecole *et al.* 2000; Perbal *et al.* 2004). Tensioning of the actin network or the other elements of the cytoskeleton in contact with the plasma membrane results in a biochemical cascade that is a source of tissue growth.

3.6.4.2.2. Extracellular matrix

The cell tissue is surrounded by a resistant pericellular medium (the cell walls) made of high-performance composite materials. This is called the extracellular matrix, which merges with the plant wall. This matrix is composed of cellulose, hemicellulose, pectin and lignin. Lignification is an impregnation of the wall by lignin (a polyphenol polymer) which increases its resistance, rigidity and hydrophobicity and accompanies cell death.

3.6.4.2.3. Summary

We can find similarities between the plant and animal kingdoms: mechanical loading is perceived by the cell membrane by the intermediary of mechanoreceptors, transmitted by the extracellular matrix to the intracellular cytoskeleton according to the tensegrity model, which is translated by alterations to exocytosis and endocytosis processes, and results in the assembly of molecules and polymers that play a part in combating gravity.

This tensegrity concept explains the "individual" internal relay of external forces. It is defined in the architectural world as "the ability of a structure to self-stabilize by the interplay of tension and compression forces that are distributed and balanced within it".

3.6.5. *Intracellular biochemical reactions*

3.6.5.1. *Animals*

(Here we speak of animals with bony skeletons.)

Activation of osteocytes by fluid movement in the canaliculi results in a biochemical cascade with production of secondary messengers such as ATP, nitric oxide, CA^{2+} and prostaglandins (Klein-Nulend *et al.* 1995; Cherian *et al.* 2005; Genetos *et al.* 2007; Lu *et al.* 2012). Cyclo-oxygenase, which participates in prostaglandin metabolism, is induced by mechanical stress, among other things, and is involved in bone remodeling and protection of cardiac function (Oshima *et al.* 2005).

3.6.5.2. *Plants*

After sedimentation of the amyloplasts, ion messengers are activated, of which calcium is the most important (Plieth and Trewavas 2002). Proteins are also involved, some of which are associated with the actin skeleton. The consequence is the production of auxin, a major growth factor, which controls the reorientation of roots towards the soil, in the direction of gravity (Perbal and Driss-Ecole 2003), as well as curvature of the branches in reaction to their weight, among other things (Figure 3.29a and b).

a) b)

Figure 3.29. *Reaction curvature of branches along the vertical axis. a) A gymnosperm (conifer). b) An angiosperm (flowering plant) (Cyrille Cazeau)*

3.6.5.3. *In total*

In both cases, a biochemical cascade is triggered, which will not be described here, but whose outcome is tissue formation.

3.6.6. *Translation by physical elements*

3.6.6.1. *Physical properties*

3.6.6.1.1. Young's modulus (E)

This is a constant property of a material that relates its deformation to the effect of physical stress. The stress applied (sigma = σ), either in compression or tension, is the relationship between an applied force (F) and the area (A) of the material's cross-section, in units of pressure (MPa):

$$\sigma = \frac{F}{S}$$

The deformation of a material ε is its observed lengthening or shortening (L) expressed relative to its initial length (Lo):

$$\varepsilon = \frac{L - L_0}{L_0}$$

The two values are related:

$$\sigma == E \times \varepsilon$$

The higher the Young's modulus (E), the stronger the material and the less it will deform for a given stress. When a bone increases its cortical thickness by mineralization (which has a higher Young's modulus than cancellous bone), it also increases its ability to withstand deformation. The same applies to a branch making lignin.

3.6.6.1.2. Second moment of area (Iz;) (SMA)

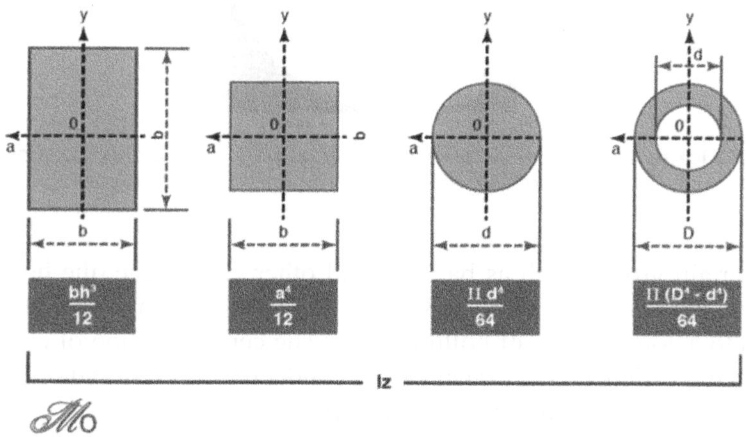

Figure 3.30. *Calculation of the second moment of area (SMA) for basic shapes, representing the object's resistance to loading based on its geometry. The SMA of the hollow tube on the left is not vastly different than the SMA of the full tube on the right, although it requires significantly less matter (source: Cazeau 2018, Maitrise Orthopédique)*

This is a geometric property of an area (Iz;) that reflects how its points are distributed relative to an arbitrary axis (Iz, horizontal here) that passes through the center of gravity. The basic formulas are shown in Figure 3.30, with the SI units being meters to the fourth power (m^4). The higher an object's SMA, the more the resistance attributed to its geometry, independent of its Young's modulus, i.e. independent of the material it is made up of.

The SMA of a full cylindrical beam is like that of a hollow cylinder. The difference is that the hollow cylinder does not require anywhere as much material to be created. In this way, the geometry of bones resembles that of the exoskeleton of arthropods (crustacean, insects and others) (Figure 3.31), and that of plants.

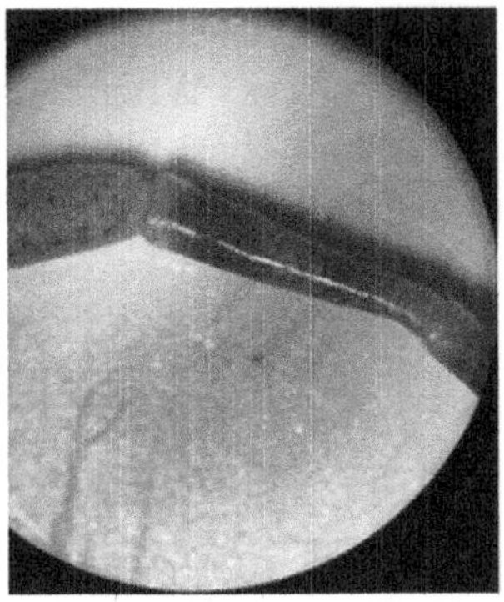

Figure 3.31. *Exoskeleton of an arthropod leg (here a shrimp); the geometry is like that of Figure 2.14a and b (Cyrille Cazeau)*

In certain groups such as bamboo and other Gramineae, the hollow tube structure is obvious (Figure 3.38a–c). For trees, its morphology can be likened to a combination of hollow tubes. The central portion of each tube is composed of aligned cells organized in conductive vessels and/or peripheral fibers, depending on the group. These fibers make up the rigid peripheral portion of tubes, composed of cellulose, pectin, lignin, etc. As the tree grows, lignification with cell death makes up the middle wood, while the activity and organization described previously continue in the peripheral portion – the cambium – which is responsible for the trunk's increase in diameter.

3.6.6.1.3. Calculation of deflection

What happens when a branch or a bone is subject to a moment? A moment is the force multiplied by the distance to its application point, in N×m. The material bends, over a distance called deflection (Figure 3.32). The amplitude of this deflection, no matter whether the beam is a bone or plant, is calculated using the SMA and Young's modulus)

$$Deflection = \frac{F \times L}{C \times E \times I_Z}$$

where:

– F: force placed on beam (weight, etc.);

– L: length of beam;

– C: constant (dimensionless);

– E: Young's modulus;

– I_Z: SMA in the Z axis.

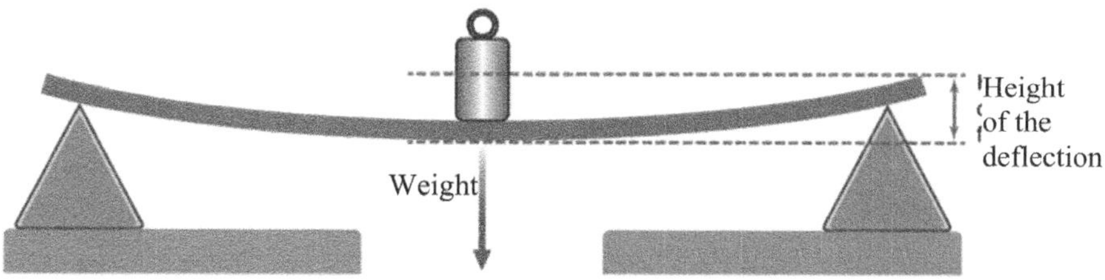

Figure 3.32. *Deflection of a beam under the action of weight; it depends on the nature of the material (Young's modulus) and the geometry of the object (second moment of area) (Cazeau 2018, Maitrise Orthopédique)*

The magnitude of a beam's deflection is proportional to the force applied, such as weight or foliage stuck in the wind. There is less deflection when the resistance to deformation is high (Young's modulus) and/or the geometry of the cross-section of the beam withstands it – bone or branch – (SMA).

The living organism (animal or plant) is able to produce matter to withstand the forces that it is subjected to by modifying the quality and orientation of its materials to optimize Young's modulus and of its geometry to increase the SMA.

3.6.6.2. *Elements modifying Young's modulus*

3.6.6.2.1. Hydroxyapatite and animal collagen

Mechanical loading provokes an increase in bone remodeling, although excessive loading can cause overload injuries (Volkmann *et al.* 1999; Robling *et al.* 2006). Conversely, the absence of stresses leads to rapid loss of bone mass like in astronauts in space or bed-ridden patients. One of the evolutionary responses in the animal kingdom was to make the skeleton more rigid by mineralizing it to enhance survival and diversification, with mechanical loading being a powerful stimulus.

Unstretchable collagen fibers mainly withstand tensile forces, while hydroxyapatite withstands the compressive ones. Thus, it is logical to see that mineralization occurs where the mechanical loads are the highest (Figure 3.33), so as to locally modify Young's modulus (Volkmann *et al.* 1999).

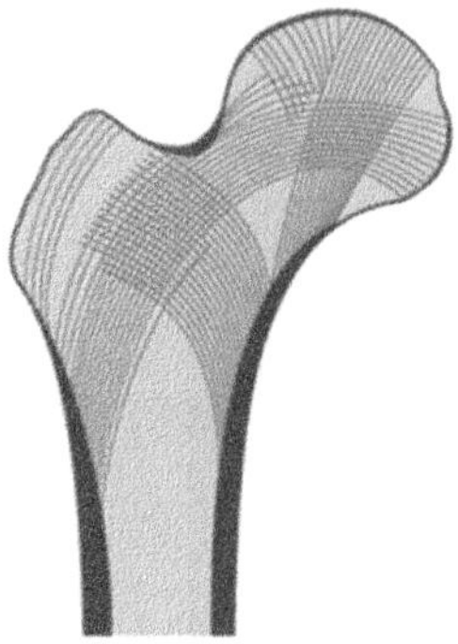

Figure 3.33. *Arrangement of cancellous bone trabeculae in the femoral neck as a function of the type and direction of stress (Cazeau 2018, Maitrise Orthopédique)*

These modifications occur in the osteon, which is a general unitary structure present in cortical bone. It is composed of osteocytes surrounded by bone lamellae and communicating in networks. The osteon is centered by the Haversian canals where nerves and blood vessels pass through.

3.6.6.2.2. Plant-based composite material development

One of the responses to mechanical changes in the environment (weight, force, wind, etc.) is the stiffening of the plant body by increasing hydrostatic pressure through hyperosmolarity, present since the appearance of the first land plants (Sachs 1868).

The other mechanism – the one we are interested in here – is the dynamic change in the type of materials, increasing Young's modulus. A plant cell wall consists of cellulose and lignin, in combination with xyloglucans (hemicellulose) and pectin, forming an extremely robust composite material. This structure is anisotropic, meaning that its mechanical properties vary by direction, which is how plants adapt to these stresses. The process of reactional lignification (reaction wood) increases its resistance. Moreover, the new wood on the periphery is highly adherent to old wood. Given its larger size, it generates a reserve of mechanical energy in the form of internal forces. Cellulose is better at resisting tensile forces, like collagen. Lignin resists compressive forces, like hydroxyapatite, providing a support function thanks to its mechanical properties and conduction function due to the hydrophobic nature of lignin. Histologically (Scurfield 1973), there are modifications in the orientation of cellulose fibers, anatomy and number of vessels (tracheids) along with lignin deposits.

Depending on the location of the mechanical stress, the tree can produce wood with an appropriate cellulose/lignin distribution. It is logical to observe that lignification always occurs where the mechanical loading is the highest, like for bone mineralization.

These products are recyclable, released outside the cell by organelles called endosomes, demonstrating the identical dynamic character of the plant or bone cell.

3.6.6.2.3. Change in trabecular bone orientation

Changes in curvature are detected inside the cancellous bone. The bone is composed of an external cortex that can withstand flexion and torsion, that is minimally elastic and that turns over more slowly than trabecular bone. The latter consists of cancellous bone, which makes up 80% of its area, but only 20% of its mass. It turns over very quickly, is not extraordinarily strong but is highly elastic.

In 1892, Julius Wolff (Frost 2003; Brand 2010; Von Meyer 2011) described the orientation of the trabeculations of cancellous bone (trabecular bone) and observed how they changed their orientation based on the mechanical stresses, in an era when X-rays had not yet been discovered. He proposed that bones change their external shape and internal architecture in response to these stresses, based on the works of the anatomist George Hermann von Meyer and the engineer Karl Culmann. The direction of the trabeculations in the cancellous bone of the proximal femur is aligned with the direction of the maximum stresses (Figure 3.34), like for the spine, tibia and calcaneus, describing its anisotropy. This made the cancellous bone stronger.

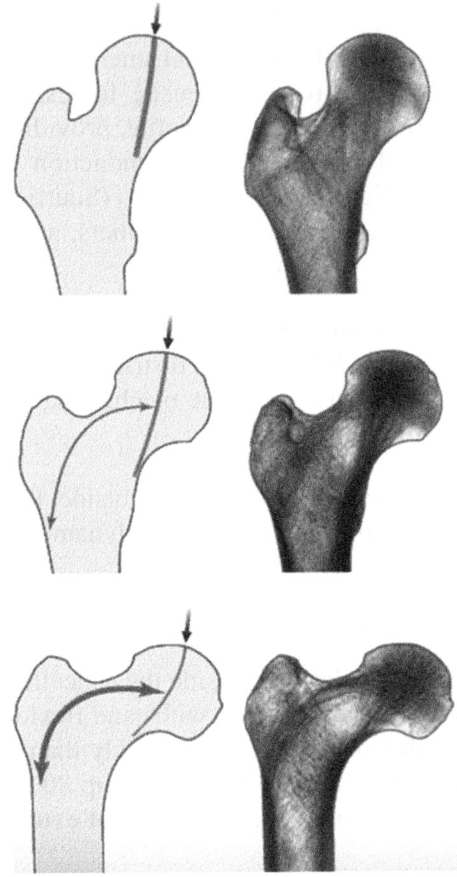

Figure 3.34. *The arrangement of bone trabeculae depends on the direction of stress according to Wolff. Diagram showing coxa valgus (top), normal femoral neck (middle) and coxa vara (bottom) (Cazeau 2018, Maitrise Orthopédique). For a color version of the figure, see www.iste.co.uk/cazeau/foot.zip*

This can be explained mechanically: when the bone bears a load that is not uniformly distributed, and/or it is already asymmetric (with a concavity and convexity), it will be subjected to excessive compression on the concave side and traction on the opposite side. The muscles mainly oppose traction forces, with the bone being responsible for withstanding compressive loads. In a simplified way, when a load is placed on a console or a cantilevered beam (Figure 3.35), it causes vertical compression forces where it is applied, and horizontal tensile forces where it is attached (wall, trunk, hip), making it the weak point.

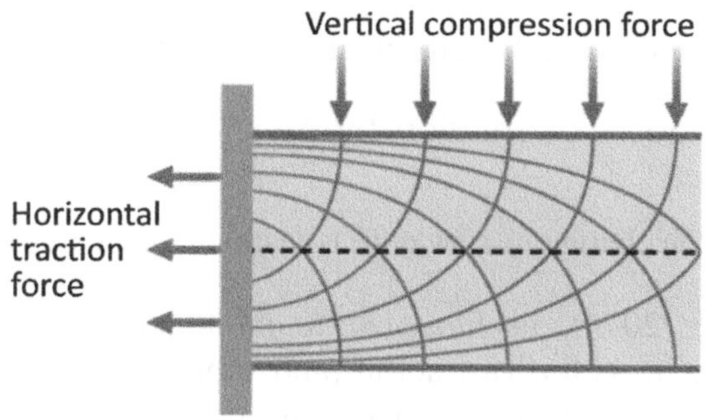

Figure 3.35. *Arrangement of vertical compressive and horizontal tensile forces on a cantilever bending system such as a large tree branch (Cazeau 2018, Maitrise Orthopédique). For a color version of the figure, see www.iste.co.uk/cazeau/foot.zip*

The appearance of a curvature with verticalization helps to suppress most of the tensile forces and to maintain the compressive forces. Qualitatively, the bone areas subjected to the most tensile loading have more collagen fibers in the axis of loading, since collagen is better at resisting traction. In compressive areas, the hydroxyapatite deposits that withstand compression are more numerous, with the collagen being arranged perpendicular to it. These elements can also be seen experimentally in animals when we alter the type of stress, evidence that this phenomenon is not solely dependent on genes. These two examples of anisotropic configuration illustrate how strength increases to adapt to the loading direction without increasing the mass cost.

Julius Wolff's work has been confirmed by recent medical archeology studies on bone ultrastructure (Gourrier and Reich 2015). Collagen is studied in several ways, from analyzing its fibrillar structure using polarized microscopy to macromolecular analysis using FTIR/Raman spectroscopy. These methods make it possible to differentiate the rapid bone formation mode through remodeling due to small mechanical stresses where fibrils are deposited anarchically (fibrous tissue) from the remodeling scenario where bone formation is slower, highly directional and influenced by the mechanical loads placed on the bone.

In the first case, the bone's response to the stresses is isotropic, while in the second case it is anisotropic, and completely adapted to the direction of the loads being applied. New analysis methods have shown that remodeling induced by loading impacts the entire bone, from its macroscopic shape to the molecular structure of the collagen and atomic structure of hydroxyapatite.

3.6.6.2.4. Change in orientation by bending of the branch: reaction wood

Unlike bone, the change in curvature is visible in the external architecture.

Francis Hallé (2014), in *Éloge de la plante*, describes the tree as an architectural system with a vast aerial (and subterranean) surface resting on an axis (the trunk) in very small proportions, composed of a small amount of living tissue covered with dead wood, the necromass (Wilson and Archer 1979). Macroscopically formed from a trunk with supporting branches considered as cantilevered, it must expose a maximum surface area to concentrate energy through photosynthesis. These branches are subjected to bending stresses due to their own weight and wind resistance. The effect of wind on foliage is complex and is a matter of aerodynamics and fluid mechanics. It can be understood in a simplified way as the effect of a moment of force. These two forces will only increase with growth.

Indeed, weight increases by the addition and growth of new branches, and wind uptake increases by widening the crown (Moulia *et al.* 2006; Niklas *et al.* 2006; Hallé and Lieutaghi 2008).

Due to the anisotropic nature of the composite material that makes up the extracellular matrix, the plant, through a change in curvature, exposes the component that best resists compression, lignin. This phenomenon also concerns the second moment of area. Furthermore, the radii of the wood and the growth rings follow the main directions of mechanical stress, allowing us to draw an analogy with "Wolff's law" described for bone (Mattheck and Bethge 1998).

3.6.6.3. *Elements modifying the second moment of area*

3.6.6.3.1. Increase in diameter

The increase in the bone diameter due to stress occurs in many situations: skeleton of children subjected to hard manual labor such as mining, and mechanical load transfer to the lateral rays in cases of functional insufficiency of the first ray of the forefoot shown in Figure 3.36 as hyperostosis of the second metatarsal.

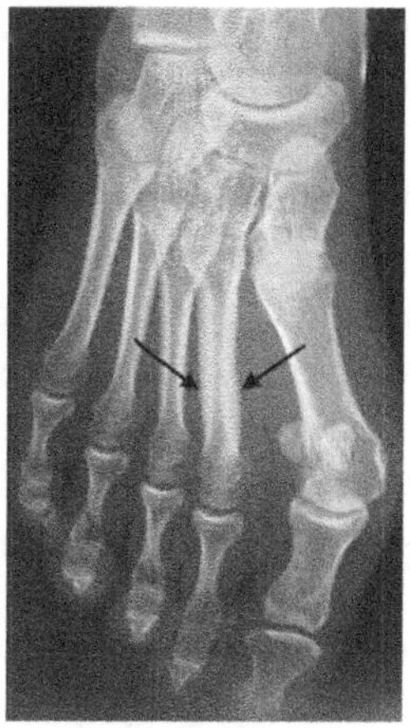

Figure 3.36. *Hyperostosis of the cortical surfaces of the second metatarsal is related to excessive stress due to load transfer from metatarsus varus*

The same goes for a fracture healing callus. Figure 3.37 shows the surgical correction of hallux valgus with percutaneous elevation osteotomies of the lateral rays that are not fixed. A bone callus appears early on and quickly evolves because of the immediate full weight bearing.

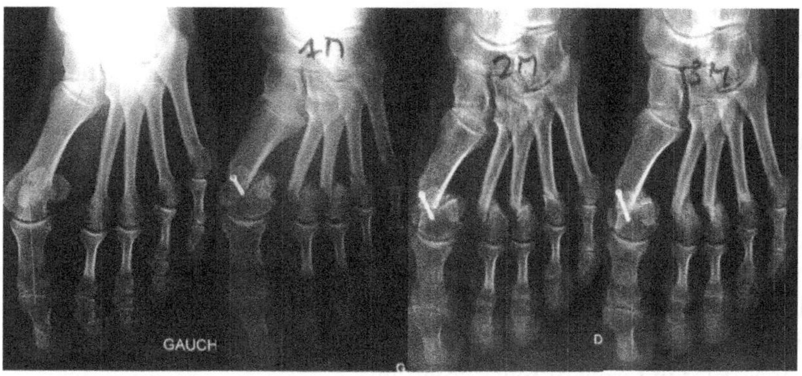

Figure 3.37. *Early onset and rapid development of bony callus after osteotomy of the first four metatarsals that were loaded immediately*

The phenomenon was described extensively by Robling *et al.* (2006). Thickening of the walls of long bones by appositional growth increases the SMA, and thereby the ability to withstand flexion and torsion.

The circular geometry helps to better counter the flexion moments in all axes (Thompson *et al.* 2009).

On the contrary, lack of loading in a growing child leads to long but small-diameter bones with lower round SMA, like a young tree that was sheltered from the wind during its growth. Experiments in dogs show bone loss with expansion of the medullary canal, predominately in the distal leg bones, the ones that are typically overloaded, close to the ground. Healthy volunteers that were confined to their beds for 17 weeks had significant bone loss, mainly in the calcaneus, lumbar spine, and proximal femur.

The impact of loading is also evident by bone overuse injuries. Their geometry and corticocancellous topography depend on the intensity and direction of the stresses (Carter and Hayes 1977; Schaffler *et al.* 1995; Norman and Wang 1997; Boyce *et al.* 1998), with the injuries triggering new repair processes as an attempt to adapt.

In botany, during growth, there is an increase in mass and lift to the wind, resulting in an increase in the diameter of the trunk and its branches. This increases the SMA by increasing the outer diameter.

3.6.6.3.2. Compromise between hollow and solid tubes

Optimization between mass and strength is embodied by the bone being shaped as a hollow tube, which allows for large reduction in mass creation for a lower reduction in SMA. The optimal ratio between the outer and inner diameter is 1:0.9 (Alexander 1996).

In the plant kingdom, stiffening occurs mainly at the periphery of the plant in an obvious manner, as in the case of Gramineae, i.e. more discreetly observable (Volkmann and Baluška 2006) around the vessels of the conducting tissues. The hollow tube geometry relative to the full tube is efficient according to the calculations of the SMA, which is slightly reduced, and saves a lot in terms of mass creation.

In both cases, the hollow tube structure is comparable to the morphology of the exoskeleton of arthropods (insects, crustaceans, etc.), or a slice of bamboo or a long bone (Figure 3.38).

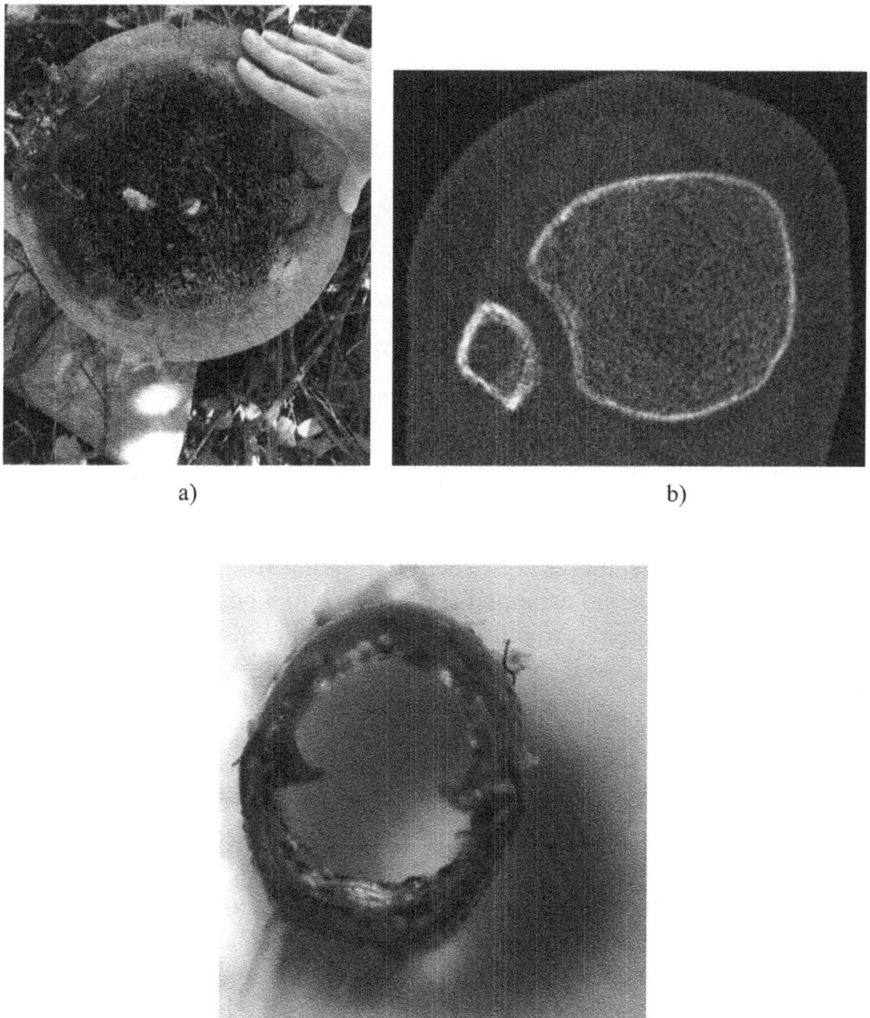

a) b) c)

Figure 3.38. *Hollow tube structure clearly visible on cross-sections: a) in bamboo, b) in long bone, c) in an arthropod leg, optimizing the second moment of area relative to the low cost of material creation (Cyrille Cazeau)*

3.6.6.3.3. Change in curvature

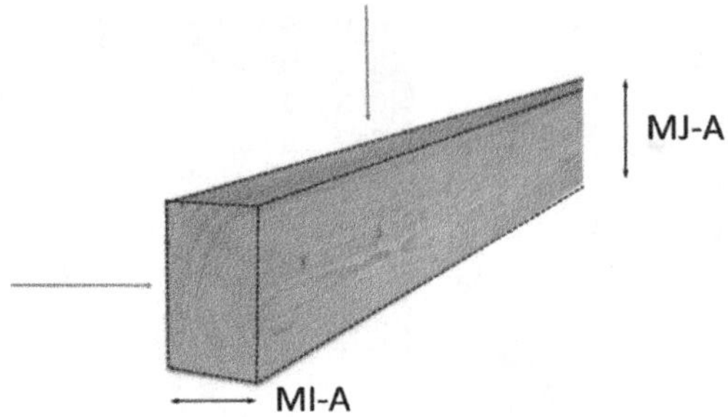

Figure 3.39. *The second moment of area is much higher when the stress is parallel to the major axis (MJ-A) of the object (MI-A; minor axis); thus, vertical straightening of the branches increases the second moment of area (Cazeau 2018, Maitrise Orthopédique)*

In schematic terms, this is equivalent to moving from studying the second moment of area of a beam arranged flat, relative to a beam of the same width and height but arranged vertically at its largest dimension (Figure 3.39). In this example, the second moment of area of the same beam arranged vertically is ten times greater.

The direction of the stress being vertical and fixed, the solution is to oppose the branch's verticalization by creating a curvature, thanks to reaction wood. The tree comes up with its own solution to resist, to compensate for the sagging of the branches and even to direct them upwards and towards the light. This is how it makes "straightening wood or reaction wood". The resistance to bending depends on the second moment of area and Young's modulus. Increasing the diameter by making lignin increases the second moment of area. The type of wood changes the bending moments and Young's modulus. The solutions are different for different classes such as gymnosperms and angiosperms. Gymnosperms (Figure 3.40a) create a compression wood, placed specifically on the underside of the branches, allowing them to straighten by pushing them towards the opposite side, and opposing their bending.

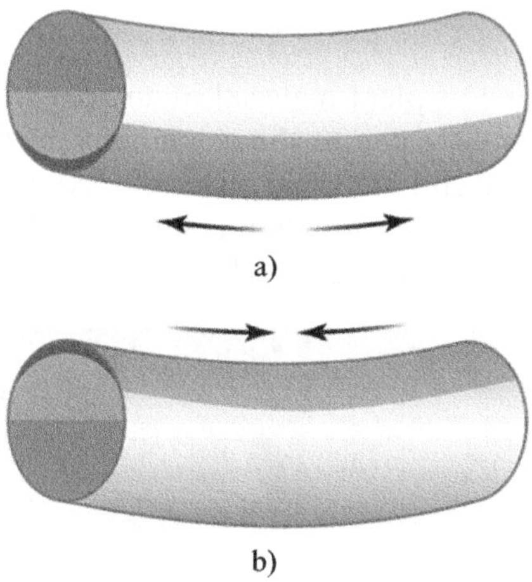

Figure 3.40. *Location of wood growth in different plant groups. a) Gymnosperms (conifers) make compression wood on the underside of branches. b) Angiosperms (flowering plants) do the opposite. This is mechanically equivalent (Cazeau 2018, Maitrise Orthopédique)*

The location in the lower convexity and the compressive strength explain the change in curvature, which occurs along the vertical axis of gravity forces as it expands longitudinally (Scurfield 1973). In angiosperms (Figure 3.40b), the solution is different but achieves the same result. They develop a reaction wood that contracts (tension wood) and resists stretching forces.

The location is on the upper side of the branches, creating superficial tension, resisting bending and straightening (Scurfield 1973). The properties are summarized in Table 3.1.

	Location of the reaction wood	**Mechanical effect**	**Resistance**
Gymnosperms (conifers)	Under side of branch	Dilation	To compression
Angiosperms (flowering plants)	Top side of branch	Contraction	To stretching

Table 3.1. *Mechanical characteristics of reaction wood under stress*

3.6.6.3.4. Ovalization of the bone's cross-section

The addition of bone or plant material in an area weakened by the repair process makes it possible – due to the asymmetry of the cross-section – to both increase the SMA and save on material. Experiments in the ulna of adult rats show that new bone formation is higher in the concave areas of the bone where the stresses are the highest (Robling *et al.* 2006) (Figure 3.41a).

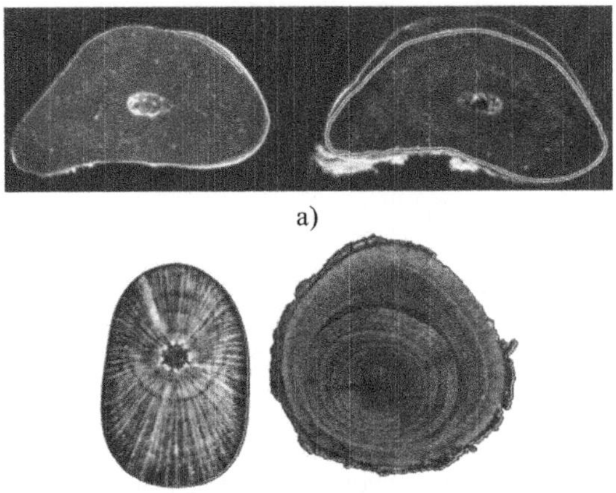

a)

b)

Figure 3.41. *a) Oval-shaped bone growth in response to stress visible on a transverse slice of the radius b) Oval growth on the underside of the branch in a gymnosperm (left and on the top side) in an angiosperm (right)*

The SMA in the experiment increased 70% by changing the bone's geometry, where it needed it the most; conversely, the 7% increase in mineralization shows that Young's modulus was minimally altered. In comparison, Figure 3.41b shows the oval-shaped growth of a gymnosperm on the left and angiosperm on the right.

3.6.6.4. *Elements modifying both parameters*

The two parameters are related when calculating the deflection, which represents the bending of a structure subjected to a bending moment. The maximum moment is applied in the middle of the beam at the "point of weakness".

In the femur, the biconvex curvature of its cortex in the middle portion of the diaphysis (Figure 3.41) exists to better withstand these moments. This feature is well known to orthopedic surgeons who carry out non-locked press-fit femoral nailing.

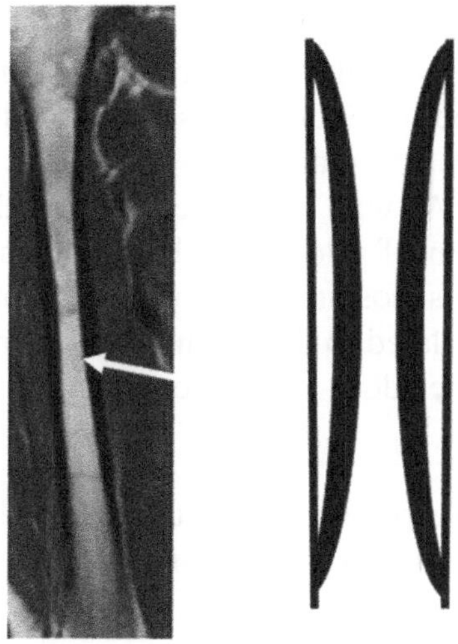

Figure 3.42. *On the anterior side of the femur, the largest amount of material is added where the bone is its weakest*

This morphology increases both Young's modulus and the SMA.

Ovalization of the cross-section during repair is another example of a mechanism impacting both physical characteristics.

3.6.6.5. *Influence of the frequency of stress application*

Other experiments have shown how the dynamics of stress application drive bone adaptation. Increased frequency increases osteogenesis considerably (Turner *et al.* 1994, 1995). Thus, it would be interesting to study the effect of walking speed (i.e. cyclic loading) in humans and variable winds on treetops. The rate of bone deformations by the intermediary of thousands of repetitive loads seems particularly effective (Xie and Liu 1999).

However, if the duration of loading is too long, the effect on formation wears out, as if the production chain became desensitized. A rest period is then needed. Given that bone is not highly innervated, the mechanoreceptors do not receive commands from a central controller.

This is like plants since they have no innervation. If a tree is felled by the wind, the shoots on the old stump (repeaters) grow vertically (Oldeman 1974) (Figure 3.18a and b). Endogenous auxin production is mainly involved in these phenomena. It stimulates the stem's growth towards the light, and the roots into the soil. As with bone, there is a local control loop which, after detection and correction of a local bending moment, stops the phenomenon. The mechanism of this postural control system on the whole plant is not known; it is probably related to the asymmetric distribution of auxin, and like poorly innervated bone, does not have a central controller (Moulia *et al.* 2006).

This is probably a rearrangement of the cytoskeleton of mechanosensitive cells, which is reversible after a refractory period.

3.6.6.6. *Summary*

There are strong analogies between plants and animals in terms of the dynamic capacity and ability to modify the structure of materials and the external shape of the "individual".

3.7. Conclusion

The remodeling and growth of human tissues vary as in bone or wood because of mechanical forces, especially those related to gravity.

While the exact mechanisms are not known, the sequencing of these phenomena is the same in plants and animals, which at least allows us to draw powerful analogies.

A mechanoreceptor is activated, transmitting the forces in the extracellular matrix to the inside of the cell to the cytoskeleton, resulting in a biochemical cascade. This translates into the production of material to repair damage, withstand stress and guide the growth necessary for survival.

Conclusion

To bring a patient as close as possible to recovery, foot surgery must focus on the recuperation of function, sometimes to the detriment of the original anatomy. The distinction between normal anatomy and useful function is best deciphered by comparative anatomy, the only one capable of making us understand the human specificity with its parameters essential to healing.

Recovery, however, must be rapid. The earliest possible restart is achieved by combining the various elements in a new concept. The concept of percutaneous surgery can be defined by the following inseparable components:

– geometric correction of the deformities in the three planes of space;

– reliable mechanical stability to allow immediate full weight bearing before any biological healing has occurred;

– minimal surgical damage to the anatomical structures being operated on;

– an anesthesia technique that will prevent the painful postoperative storm associated with conventional techniques.

References

Aiello, L. and Dean, C. (1990). *An Introduction to Human Evolutionary Anatomy*. Academic Press, London.

Alexander, R.M. (1996). *Optima for Animals*. Princeton University Press, Princeton.

Anken, R.H. (2003). Neurophysiology of developing fish at altered gravity: Background–facts–perspectives [Online]. *Advances in Space Biology and Medicine*, 9, 173–200. Available: https://doi.org/10.1016/S1569-2574(03)09007-5.

Anken, R.H. (2006). On the role of the central nervous system in regulating the mineralisation of inner-ear otoliths of fish [Online]. *Protoplasma*, 229, 205–208. Available: https://doi.org/10.1007/s00709-006-0219-6.

Aseyev, N. (2017). Adaptive changes in the vestibular system of land snail to a 30-day spaceflight and readaptation on return to earth [Online]. *Frontiers in Cellular Neuroscience*, 11. Available: https://doi.org/10.3389/fncel.2017.00348.

Ashton, N., Lewis, S.G., De Groote, I., Duffy, S.M., Bates, M., Bates, R., Hoare, P., Lewis, M., Parfitt, S.A., Peglar, S., Williams, C., and Stringer, C. (2014). Hominin footprints from early pleistocene deposits at happisburgh, UK [Online]. *PLoS ONE*, 9. Available: https://doi.org/10.1371/journal.pone.0088329.

Austin, D.W. and Leventen, E.O. (1981). A new osteotomy for hallux valgus: A horizontally directed "V" displacement osteotomy of the metatarsal head for hallux valgus and primus varus. *Clinical Orthopaedics and Related Research*, 157, 25–30.

Bacon, A.-M. (2001). *La locomotion des primates miocènes d'Afrique et d'Europe : analyse fonctionnelle des os longs du membre pelvien et systématique*. CNRS, Paris.

Badwey, T.M., Dutkowsky, J.P., Graves, S.C., and Richardson, E.G. (1997). An anatomical basis for the degree of displacement of the distal chevron osteotomy in the treatment of hallux valgus [Online]. *Foot & Ankle Society*, 18, 213–215. Available: https://doi.org/10.1177/107110079701800405.

Barouk, L.S. (2012). *Brièveté des gastrocnémiens : de l'anatomie au traitement.* Sauramps Médical, Paris.

Barouk, L.S. and Barouk, P. (2007). The scarf first metatarsal osteotomy in the correction of hallux valgus deformity [Online]. *Interactive Surgery*, 2, 2–11. Available: https://doi.org/10.1007/s11610-007-0023-9.

Barouk, L.S., Barouk, P., and Toullec, E. (2005). Brièveté des muscles gastrocnémiens et pathologie de l'avant-pied : la libération proximale chirurgicale [Online]. *Médecine et chirurgie du pied*, 21, 143–152. Available: https://doi.org/10.1007/s10243-005-0063-y.

Beaumont, A. and Cassier, P. (1997). *Biologie animale : Les cordés, anatomie comparée des vertébrés*. Dunod, Paris.

Benichou, M. (2006). Traitement chirurgical de l'hallux valgus par mini-abord. *Maîtrise Orthopédique*, 159, 4–6.

Benton, M.J. and Desbois, H. (1998). *Atlas historique des dinosaures 1 000 espèces, 160 millions d'années*. Autrement, Paris.

Bérillon, G. (2000). *Le pied des hominoïdes miocènes et des hominidés fossiles : Architecture, locomotion, évolution.* CNRS, Paris.

Bérillon, G. and Marchal, F. (2005). Les multiples bipédies. *Pour la science*, 330, 76–83.

Besova, N.V., Savel'ev, S.V., and Chernikov, V.P. (1993). The effect of weightlessness on amphibians. The skeleton and mineral metabolism. *Bulletin of Experimental Biology and Medicine*, 116, 90–92.

Bhatia, D., Myerson, M.S., Curtis, M.J., Cunningham, B.W., and Jinnah, R.H. (1994). Anatomical restraints to dislocation of the second metatarsophalangeal joint and assessment of a repair technique [Online]. *Journal of Bone and Joint Surgery*, 76, 1371–1375. Available: https://doi.org/10.2106/00004623-199409000-00012.

Bouisset, S. (2002). *Biomécanique et physiologie du mouvement*. Masson, Paris.

Boyce, T.M., Fyhrie, D.P., Glotkowski, M.C., Radin, E.L., and Schaffler, M.B. (1998). Damage type and strain mode associations in human compact bone bending fatigue [Online]. *Journal of Orthopaedic Research*, 16, 322–329. Available: https://doi.org/10.1002/jor.1100160308.

Brand, R.A. (2010). Biographical sketch: Julius Wolff, 1836–1902 [Online]. *Clinical Orthopaedics and Related Research*, 468, 1047–1049. Available: https://doi.org/10.1007/s11999-010-1258-z.

Buffetaut, E. (1998). *Histoire de la paléontologie*. Presses universitaires de France, Paris.

Burger, E.H. and Klein-Nulend, J. (1999). Mechanotransduction in bone–role of the lacuno-canalicular network [Online]. *The FASEB Journal*, 13, 101–112. Available: https://doi.org/10.1096/fasebj.13.9001.s101.

Burr, D.B., Martin, R.B., Schaffler, M.B., and Radin, E.L. (1985). Bone remodeling in response to in vivo fatigue microdamage. *Journal of Biomechanics*, 18, 189–200.

Carter, D.R. and Hayes, W.C. (1977). Compact bone fatigue damage: A microscopic examination. *Clin Orthop Relat Res*, 1977 (127), 265–74.

Cavagna, G.A., Dusman, B., and Margaria, R. (1968). Positive work done by a previously stretched muscle [Online]. *Journal of Applied Physiology*, 24, 21–32. Available: https://doi.org/10.1152/jappl.1968.24.1.21.

Cazeau, C. (2004a). Rencontre : Mariano de Prado, Christophe de Lavigne. *Maîtrise Orthopédique*, 133.

Cazeau, C. (2004b). Anatomie fonctionnelle comparée de l'articulation de la cheville. *Maîtrise Orthopédique*, 133, 1–12.

Cazeau, C. (2006). Faut-il s'intéresser à la bipédie ? Analyse anatomique et biomécanique dans diverses classes animales. Utilité pour la recherche de la phylogénie humaine. *Maîtrise Orthopédique*, 155, 12–20.

Cazeau, C. (2007a). Analyse biomécanique d'une spécificité de la bipédie humaine : la mise en flexion dorsale de la cheville à l'appui. *Maîtrise Orthopédique*, 166, 22–26.

Cazeau, C. (2007b). Chirurgie mini-invasive du pied et de la cheville. Rapport du congrès d'Arcachon. *Maîtrise Orthopédique*, 169.

Cazeau, C. (2008a). Chirurgie mixte mini-invasive de l'hallux valgus : Limites de la chirurgie percutanée. *Maîtrise Orthopédique*, 179, 12–16.

Cazeau, C. (2008b). Le geste en chirurgie orthopédique. *Revue Geste*, September, 322–339.

Cazeau, C. (2008c). La chirurgie percutanée de l'avant-pied : évolution vers la chirurgie mini-invasive. *Rhumatos*, 5(37), 169–175.

Cazeau, C. (2009a). Arthrolyse de la première articulation métatarso-phalangienne. In *Chirurgie mini-invasive et percutanée du pied*, Cazeau, C. (ed.). Sauramps Médical, Montpellier, 67–74.

Cazeau, C. (2009b). Intérêt de la biométrie dans la chirurgie assistée par ordinateur. *Biométrie Humaine et Anthropologie*, 103–105.

Cazeau, C. (2009c). Place du traitement chirurgical préventif du pied diabétique. Place de la chirurgie percutanée. *Les Entretiens de Bichat-Podologie*. Palais des Congrès, Paris.

Cazeau, C. (2011). Syndrome canalaire : de la décompression à l'exérèse. *Les Entretiens de Bichat-Podologie*. Palais des Congrès, Paris.

Cazeau, C. (2012a). L'aseptie du pied en salle d'opération. *Maîtrise Orthopédique*, 214, 14–17.

Cazeau, C. (2012b). Orteils en griffes : de la biomécanique à la chirurgie. *Les Entretiens de Bichat-Podologie*. Palais des Congrès, Paris.

Cazeau, C. (2013a). Syndrome du deuxième rayon. *Les Entretiens de Bichat-Podologie*. Palais des Congrès, Paris.

Cazeau, C. (2013b). Biomécanique des pathologies de contraintes de l'avant-pied et du médio-pied. *Les Entretiens de Bichat-Podologie*. Palais des Congrès, Paris.

Cazeau, C. (2014). Hyper-appui et pathologie sésamoïdienne. *Les Entretiens de Bichat-Podologie*. Palais des Congrès, Paris.

Cazeau, C. (ed.) (2015). Groupe de recherche et d'étude en chirurgie mini-invasive du pied et de la cheville. *Chirurgie mini-invasive et percutanée du pied.* Sauramps Médical, Montpellier.

Cazeau, C. (2016). Étude biomécanique de la plaque plantaire. *Maîtrise Orthopédique*, 252(1), 22–23.

Cazeau, C. (2018). Gravité et croissance. *Maîtrise Orthopédique*, 278(1), 14–21.

Cazeau, C. and Chevalier, T. (2014). Le syndrome du deuxième rayon. *Maîtrise Orthopédique*, 238.

Cazeau, C. and Chevalier, T. (2016). Anatomie comparée du deuxième métatarsien chez les primates actuels et fossiles [Online]. Available: https://www.researchgate.net/publication/312093345_2016_Anatomie_comparee_du_deuxieme_metatarsien_chez_les_primates_actuels_et_fossiles.

Cazeau, C. and Diebold, P. (2017). L'ostéotomie épiphyso-métaphysaire distale en chevron. *Maîtrise Orthopédique*, 264(1), 4–9.

Cazeau, C. and Stiglitz, Y. (2012). Analyse des conséquences biomécaniques de la brièveté du gastrocnémien sur l'avant-pied [Online]. Available: https://www.researchgate.net/profile/Yves_Stiglitz2/publication/268036301_Analisis_de_las_consecuencias_biomecanicas_del_gastrocnemio_corto_en_le_antepie/links/5c152bf14585157ac1c562e3/Analisis-de-las-consecuencias-biomecanicas-del-gastrocnemio-corto-en-le-antepie.pdf?origin=publication_detail.

Cazeau, C. and Stiglitz, Y. (2014). Effects of gastrocnemius tightness on forefoot during gait [Online]. *Foot and Ankle Clinics*, 19, 649–657. Available: https://doi.org/10.1016/j.fcl.2014.08.003.

Cazeau, C. and Stiglitz, Y. (2018). Minimally invasive and percutaneous surgery of the forefoot current techniques in 2018 [Online]. *European Journal of Orthopaedic Surgery & Traumatology*, 28, 819–837. Available: https://doi.org/10.1007/s00590-018-2137-7.

Cazeau, C. and Stiglitz, Y. (2019). Maladie ou syndrome de Morton [Online]. Available: https://www.em-consulte.com/article/1062632/maladie-ou-syndrome-de-morton.

Chalayon, O., Chertman, C., Guss, A.D., Saltzman, C.L., Nickisch, F., and Bachus, K.N. (2013). Role of plantar plate and surgical reconstruction techniques on static stability of lesser metatarsophalangeal joints: A biomechanical study [Online]. *Foot Ankle International*, 34, 1436–1442. Available: https://doi.org/10.1177/1071100713491728.

Cherian, P.P., Siller-Jackson, A.J., Gu, S., Wang, X., Bonewald, L.F., Sprague, E., and Jiang, J.X. (2005). Mechanical strain opens connexin 43 hemichannels in osteocytes: A novel mechanism for the release of prostaglandin [Online]. *Molecular Biology of the Cell*, 16, 3100–3106. Available: https://doi.org/10.1091/mbc.e04-10-0912.

Colombier, J.-A. (2015). *Chirurgie du Pied et de la Cheville en Mode Ambulatoire*. Elsevier Masson, Paris.

Continenza, B. (2009). *Darwin : L'arbre de Vie*. Belin, Paris.

Coppens, Y. and Picq, P.G. (eds) (2001). *Aux origines de l'humanité*. Fayard, Paris.

Czeisler, C.A., Chiasera, A.J., and Duffy, J.F. (1991). Research on sleep, circadian rhythms and aging: Applications to manned spaceflight [Online]. *Experimental Gerontology*, 26, 217–232. Available: https://doi.org/10.1016/0531-5565(91)90014-D.

D'Août, K., Aerts, P., De Clercq, D., De Meester, K., and Van Elsacker, L. (2002). Segment and joint angles of hind limb during bipedal and quadrupedal walking of the bonobo (*Pan paniscus*): Bipedal and Quadrupedal Walking in Bonobos [Online]. *American Journal of Physical Anthropology*, 119, 37–51. Available: https://doi.org/10.1002/ajpa.10112.

De Lavigne, C. (ed.) (2004). Le traitement de l'hallux valgus par technique mini-invasive. *Maîtrise Orthopédique*, 133.

Deloison, Y. (2004). *La Préhistoire du Piéton : Essai sur les Nouvelles Origines de L'homme*. Plon, Paris.

DeSilva, J.M., Proctor, D.J., and Zipfel, B. (2012). A complete second metatarsal (StW 89) from sterkfontein member 4 [Online]. *South African Medical Journal*, 63, 487–496. Available: https://doi.org/10.1016/j.jhevol.2012.05.010.

Di, S.M., Qian, A.R., Qu, L.N., Zhang, W., Wang, Z., Ding, C., Li, Y.H., Ren, H.G., and Shang, P. (2011). Graviresponses of osteocytes under altered gravity [Online]. *Advances in Space Research*, 48, 1161–1166. Available: https://doi.org/10.1016/j.asr.2011.05.030.

Diebold, P. and Cazeau, C. (2007). Le chevron (ostéotomie épiphyso-métaphysaire distale en V du premier métatarsien). *Maîtrise Orthopédique*, 169.

DiGiovanni, C.W. and Langer, P. (2007). The role of isolated gastrocnemius and combined Achilles contractures in the flatfoot [Online]. *Foot and Ankle Clinics*, 12, 363–379. Available: https://doi.org/10.1016/j.fcl.2007.03.005.

DiGiovanni, C.W., Kuo, R., Tejwani, N., Price, R., Hansen, S.T., Cziernecki, J., and Sangeorzan, B.J. (2002). Isolated gastrocnemius tightness [Online]. *Journal of Bone and Joint Surgery*, 84, 962–970. Available: https://doi.org/10.2106/00004623-200206000-00010.

Doran, D.M. (1997). Ontogeny of locomotion in mountain gorillas and chimpanzees [Online]. *Journal of Human Evolution*, 32, 323–344. Available: https://doi.org/10.1006/jhev.1996.0095.

Doran, D.M., Goodall, J., and Itani, J. (1996). Comparative positional behavior of the African apes [Online]. In *Great Ape Societies*, McGrew, W.C., Marchant, L.F., and Nishida, T. (eds). Cambridge University Press, Cambridge, 213–224. Available: https://doi.org/10.1017/CBO9780511752414.018.

Drapeau, M.S.M. and Harmon, E.H. (2013). Metatarsal torsion in monkeys, apes, humans and australopiths [Online]. *Journal of Human Evolution*, 64, 93–108. Available: https://doi.org/10.1016/j.jhevol.2012.10.008.

Driss-Ecole, D., Jeune, B., Prouteau, M., Julianus, P., and Perbal, G. (2000). Lentil root statoliths reach a stable state in microgravity [Online]. *Planta*, 211, 396–405. Available: https://doi.org/10.1007/s004250000298.

Ducroquet, R.J. and Ducroquet, P. (1964). *La Marche et les Boiteries*. Masson, Paris.

Eberhart, H.D., Inman, V.T., and Bresler, B. (1954). The principal elements in human locomotion. In *Human Limbs and their Substitutes*, Klopsteg, P.E., Wilson, P.D. (eds). McGraw-Hill, New York, 437–471.

Edelmann, E., Anken, R.H., and Rahmann, H. (2004). Swimming behaviour and calcium incorporation into inner ear otoliths of fish after vestibular nerve transection [Online]. *Advances in Space Research*, 33, 1390–1394. Available: https://doi.org/10.1016/j.asr.2003.09.042.

Evans, F.G. (1976). Mechanical properties and histology of cortical bone from younger and older men [Online]. *The Anatomical Record*, 185, 1–11. Available: https://doi.org/10.1002/ar.1091850102.

Fitzelle, K.J. and Kiss, J.Z. (2001). Restoration of gravitropic sensitivity in starch-deficient mutants of Arabidopsis by hypergravity [Online]. *Journal of Experimental Botany*, 52, 265–275. Available: https://doi.org/10.1093/jexbot/52.355.265.

Fleagle, J.G. (2013). *Primate Adaptation and Evolution*. Elsevier/Academic Press, Amsterdam/Boston.

Frost, H.M. (2003). Bone's mechanostat: a 2003 update [Online]. *Anatomical Record: Advances in Integrative Anatomy and evolutionary Biology*, 275, 1081–1101. Available: https://doi.org/10.1002/ar.a.10119.

Fukumoto, S. (2008). PTH and mechanical stress [Online]. *Clinical Calcium*, 18, 1327–1331. Available: https://doi.org/CliCa080913271331.

Fuller, P.M., Warden, C.H., Barry, S.J., and Fuller, C.A. (2000). Effects of 2-G exposure on temperature regulation, circadian rhythms, and adiposity in UCP2/3 transgenic mice [Online]. *Journal of Applied Physiology*, 89, 1491–1498. Available: https://doi.org/10.1152/jappl.2000.89.4.1491.

Gadomski, B.C., McGilvray, K.C., Easley, J.T., Palmer, R.H., Santoni, B.G., and Puttlitz, C.M. (2014). Partial gravity unloading inhibits bone healing responses in a large animal model [Online]. *Journal of Biomechanics*, 47, 2836–2842. Available: https://doi.org/10.1016/j.jbiomech.2014.07.031.

Gambrill, M.R., Ledney, G.D., and MacVittie, T.J. (1976). Mitigation of graft-versus-host disease in lethally irradiated mice grafted with spleen cells adherent to glass beads. *Transplantation*, 21, 247–254.

Gauthier, G. (1979). Thomas Morton's disease: a nerve entrapment syndrome. A new surgical technique. *Clinical Orthopaedics and Related Research*, 142, 90–92.

Genetos, D.C., Kephart, C.J., Zhang, Y., Yellowley, C.E., and Donahue, H.J. (2007). Oscillating fluid flow activation of gap junction hemichannels induces ATP release from MLO-Y4 osteocytes [Online]. *Journal of Cellular Physiology*, 212, 207–214. Available: https://doi.org/10.1002/jcp.21021.

Gnyubkin, V., Guignandon, A., Laroche, N., Vanden-Bossche, A., Normand, M., Lafage-Proust, M.-H., and Vico, L. (2015). Effects of chronic hypergravity: From adaptive to deleterious responses in growing mouse skeleton [Online]. *Journal of Applied Physiology*, 119, 908–917. Available: https://doi.org/10.1152/japplphysiol.00364.2015.

Gould, S.J. (ed.) (2001). *The Book of Life*. Norton, New York.

Gourrier, A. and Reiche, I. (2015). L'os : morphologie, structure et composition chimique. In *Messages d'os : archéométrie du squelette animal et humain*, Balasse, M. (ed.). Éditions des Archives Contemporaines, Paris.

Grassé, P.-P., Doumenc, D., Érard, C., and Meunier, F. (1996). *Vertébrés*. Masson, Paris.

Gray, H. and Clemente, C.D. (1985). *Anatomy of the Human Body*. Lea and Febiger, Philadelphia.

Groupe de recherche et d'évaluation en chirurgie mini-invasive du pied et de la cheville (2010). *Arthroscopies et Endoscopies de la Cheville et du Pied*. Sauramps Médical, Montpellier.

Haile-Selassie, Y., Saylor, B.Z., Deino, A., Levin, N.E., Alene, M., and Latimer, B.M. (2012). A new hominin foot from Ethiopia shows multiple Pliocene bipedal adaptations [Online]. *Nature*, 483, 565–569. Available: https://doi.org/10.1038/nature10922.

Hallé, F. (1999). *Éloge de la Plante : Pour une Nouvelle Biologie*. Le Seuil, Paris.

Hallé, F. and Lieutaghi, P. (eds) (2008). *Aux Origines des Plantes*. Fayard, Paris.

Hama, N., Tsuchida, Y., and Takahata, M. (2007). Behavioral context-dependent modulation of descending statocyst pathways during free walking, as revealed by optical telemetry in crayfish [Online]. *Journal of Experimental Biology*, 210, 2199–2211. Available: https://doi.org/10.1242/jeb.002865.

Han, Y., Cowin, S.C., Schaffler, M.B., and Weinbaum, S. (2004). Mechanotransduction and strain amplification in osteocyte cell processes [Online]. *Proceedings of the National Academy of Sciences of the United States of America*, 101, 16689–16694. Available: https://doi.org/10.1073/pnas.0407429101.

Hansen, J.T. and Netter, F.H. (2014). *Netter's Anatomy Flash Cards*. Saunders, Philadelphia.

Hartenberger, J.-L. (2001). *Une Brève Histoire des Mammifères : Bréviaire de Mammalogie*. Belin, Paris.

Heer, M., Kamps, N., Biener, C., Korr, C., Boerger, A., Zittermann, A., Stehle, P., and Drummer, C. (1999). Calcium metabolism in microgravity. *European Journal of Medical Research*, 4, 357–360.

Hérisson, C., Aboukrat, P., Rodineau, J., and Alomène, B. (2005). *La Rétraction du Triceps Sural : Conséquences Podologiques*. Sauramps Médical, Montpellier.

Hill, A.V. (1938). The heat of shortening and the dynamic constants of muscle [Online]. *Proceedings of the Royal Society of London Series B – Biological Sciences*, 126, 136–195. Available: https://doi.org/10.1098/rspb.1938.0050.

Holick, M.F. (1998). Perspective on the impact of weightlessness on calcium and bone metabolism. *Bone*, 22, 105–111.

Huson, A. (1991). Functional anatomy of the foot. In *Disorders of the Foot and Ankle*, 1, Jahs, J.H. (ed.). Saunders, Philadelphia.

Ingber, D.E. (1998). The architecture of life [Online]. *Scientific American*, 278, 48–57. Available: https://doi.org/10.1038/scientificamerican0198-48.

Iolascon, G., Resmini, G., and Tarantino, U. (2013). Mechanobiology of bone [Online]. *Aging Clinical and Experimental Research*, 25, 3–7. Available: https://doi.org/10.1007/s40520-013-0101-2.

Ishay, J.S., Shimony, T., and Arcan, L. (1983). The presence of statocysts and statoliths in social wasps (hymenoptera, vespinae) [Online]. *Life Sciences*, 32, 1711–1719. Available: https://doi.org/10.1016/0024-3205(83)90833-0.

Jahss, M.H. (ed.) (1991). *Disorders of the Foot & Ankle: Medical and Surgical Management*. Saunders, Philadelphia.

Janvier, P. (ed.) (1989). *La recherche en paléontologie*. Le Seuil, Paris.

Jay, P.J. and Schaller, G.B. (1963). The mountain gorilla. Ecology and behavior [Online]. *American Journal of Physical Anthropology*, 23, 440–442. Available: https://doi.org/10.1002/ajpa.1330230421.

Jay, P.J. and Schaller, G.B. (1965). *The Mountain Gorilla. Ecology and Behavior.* University of Chicago Press, Chicago.

Johanson, M.A., Cuda, B.J., Koontz, J.E., Stell, J.C., and Abelew, T.A. (2009). Effect of stretching on ankle and knee angles and gastrocnemius activity during the stance phase of gait. *Journal of Sport Rehabilitation*, 18, 521–534.

Johnson, K.A., Cofield, R.H., and Morrey, B.F. (1979). Chevron osteotomy for hallux valgus. *Clinical Orthopaedics and Related Research*, 1979 (142), 44–7.

Johnston, R.B., Smith, J., and Daniels, T. (1994). The plantar plate of the lesser toes: An anatomical study in human cadavers [Online]. *Foot Ankle International*, 15, 276–282. Available: https://doi.org/10.1177/107110079401500508.

Kapandji, A.I. and Merle d'Aubigné, R. (1975). Physiologie articulaire. *Schémas commentés de mécanique humaine*, III. Maloine, Paris.

Kawao, N., Morita, H., Obata, K., Tamura, Y., Okumoto, K., and Kaji, H. (2016). The vestibular system is critical for the changes in muscle and bone induced by hypergravity in mice [Online]. *Physiological Reports*, 4. Available: https://doi.org/10.14814/phy2.12979.

Kelikian, H. (1966). Hallux valgus, allied deformities of the forefoot and metatarsalgia [Online]. *British Journal of Surgery*, 53. Available: https://doi.org/10.1002/bjs.1800530141.

Kelikian, A.S. and Sarrafian, S.K. (eds.) (2011). *Sarrafian's Anatomy of the Foot and Ankle: Descriptive, Topographical, Functional.* Lippincott Williams & Wilkins, Philadelphia.

Kelly, P.J. and Bronk, J.T. (1990). Venous pressure and bone formation. *Microvascular Research*, 39, 364–375.

Khan, K., Brown, J., Way, S., Vass, N., Crichton, K., Alexander, R., Baxter, A., Butler, M., and Wark, J. (1995). Overuse injuries in classical ballet [Online]. *Sports Medicine*, 19, 341–357. Available: https://doi.org/10.2165/00007256-199519050-00004.

Kharkeevitch, T.A. and Titova, L.K. (1974). Electron microscope studies of the structural organization of statocyst of arenicola marina. *Z Mikrosk Anat Forsch*, 88, 382–392.

Klein-Nulend, J., Semeins, C.M., Ajubi, N.E., Nijweide, P.J., and Burger, E.H. (1995). Pulsating fluid flow increases nitric oxide (NO) synthesis by osteocytes but not periosteal fibroblasts–correlation with prostaglandin upregulation [Online]. *Biochemical and Biophysical Research Communications*, 217, 640–648. Available: https://doi.org/10.1006/bbrc.1995.2822.

Kuttruff, J.T., DeHart, S.G., and O'Brien, M.J. (1998). 7500 years of prehistoric footwear from arnold research cave, Missouri [Online]. *Science*, 281, 72–75. Available: https://doi.org/10.1126/science.281.5373.72.

Kuznetsov, O. (1997). Magnetophoretic induction of curvature in coleoptiles and hypocotyls [Online]. *Journal of Experimental Botany*, 48, 1951–1957. Available: https://doi.org/10.1093/jexbot/48.316.1951.

Lanyon, L.E. (2009). The influence of function on the development of bone curvature. An experimental study on the rat tibia [Online]. *Journal of Zoology*, 192, 457–466. Available: https://doi.org/10.1111/j.1469-7998.1980.tb04243.x.

Laredo, J.-D. (2011). *Savoir Faire en Radiologie Ostéo-articulaire*, 13. Sauramps Médical, Montpellier.

Laredo, J.-D., Wybier, M., Bousson, V., Parlier, C., Petrover, D., and Miquel, A. (2013). *Savoir faire en radiologie ostéo-articulaire*, 15. Sauramps Médical, Montpellier.

Laurent, M.R., Dubois, V., Claessens, F., Verschueren, S.M.P., Vanderschueren, D., Gielen, E., and Jardí, F. (2016). Muscle-bone interactions: From experimental models to the clinic? A critical update [Online]. *Molecular and Cellular Endocrinology*, 432, 14–36. Available: https://doi.org/10.1016/j.mce.2015.10.017.

Leblanc, A.D., Schneider, V.S., Evans, H.J., Engelbretson, D.A., and Krebs, J.M. (1990). Bone mineral loss and recovery after 17 weeks of bed rest [Online]. *Journal of Bone and Mineral Research*, 5, 843–850. Available: https://doi.org/10.1002/jbmr.5650050807.

Lecointre, G. and Le Guyader, H. (2006). *Classification Phylogénétique du Vivant*. Belin, Paris.

Lécuru-Renous, S. (1994). *Locomotion*. Dunod, Paris.

Lee, H.-J., Chung, J.-W., Chu, I.-T., and Kim, Y.-C. (2010). Comparison of distal chevron osteotomy with and without lateral soft tissue release for the treatment of hallux valgus [Online]. *Foot Ankle International*, 31, 291–295. Available: https://doi.org/10.3113/FAI.2010.0291.

Lee, K.B., Cho, N.Y., Park, H.W., Seon, J.K., and Lee, S.H. (2015). A comparison of proximal and distal chevron osteotomy, both with lateral soft-tissue release, for moderate to severe hallux valgus in patients undergoing simultaneous bilateral correction: A prospective randomised controlled trial [Online]. *Bone & Joint Journal*, 97-B, 202–207. Available: https://doi.org/10.1302/0301-620X.97B2.34449.

Leemrijse, T. (2015). *Pathologie du Pied et de la Cheville*. Elsevier Masson, Paris.

Leitz, G., Kang, B.-H., Schoenwaelder, M.E.A., and Staehelin, L.A. (2009). Statolith sedimentation kinetics and force transduction to the cortical endoplasmic reticulum in gravity-sensing arabidopsis columella cells [Online]. *The Plant Cell*, 21, 843–860. Available: https://doi.org/10.1105/tpc.108.065052.

Lindgren, U. and Turan, I. (1983). A new operation for hallux valgus [Online]. *Clinical Orthopaedics and Related Research*, 179–183. Available: https://doi.org/10.1097/00003086-198305000-00026.

Lu, X.L., Huo, B., Park, M., and Guo, X.E. (2012). Calcium response in osteocytic networks under steady and oscillatory fluid flow [Online]. *Bone*, 51, 466–473. Available: https://doi.org/10.1016/j.bone.2012.05.021.

Maestro, M., Leemrijse, T., and Besse, J.-L. (eds.) (2009). *Association Française de Chirurgie du Pied. Journées de Spécialités*. Sauramps Médical, Montpellier.

Mattheck, C. and Bethge, K. (1998). The structural optimization of trees [Online]. *Naturwissenschaften*, 85, 1–10. Available: https://doi.org/10.1007/s001140050443.

McGrew, W.C., Marchant, L.F., and Nishida, T. (eds.) (1996). *Great Ape Societies*. Cambridge University Press, Cambridge.

Milgrom, C., Giladi, M., Stein, M., Kashtan, H., Margulies, J., Chisin, R., Steinberg, R., and Aharonson, Z. (1985). Stress fractures in military recruits. A prospective study showing an unusually high incidence [Online]. *The Journal of Bone and Joint Surgery*, 67B, 732–735. Available: https://doi.org/10.1302/0301-620X.67B5.4055871.

Mitchell, C.L., Fleming, J.L., Allen, R., Glenney, C., and Sanford, G.A. (1958). Osteotomy-bunionectomy for hallux valgus. *Journal of Bone and Joint Surgery*, 40-A, 41–600.

Moody, S.A. and Golden, C. (2000). Developmental biology research in space: Issues and directions in the era of the international space station [Online]. *Developmental Biology*, 228, 1–5. Available: https://doi.org/10.1006/dbio.2000.9907.

Morton, D.J. (1924). Evolution of the human foot, II [Online]. *American Journal of Physical Anthropology*, 7, 1–52. Available: https://doi.org/10.1002/ajpa.1330070114.

Morton, D.J. (1936). The human foot: Its evolution, physiology and functional disorders [Online]. *BJS*, 24(93). Available: https://doi.org/10.1002/bjs.1800249344.

Moulia, B., Coutand, C., and Lenne, C. (2006). Posture control and skeletal mechanical acclimation in terrestrial plants: Implications for mechanical modeling of plant architecture [Online]. *American Journal of Botany*, 93, 1477–1489. Available: https://doi.org/10.3732/ajb.93.10.1477.

Murakami, D.M. and Fuller, C.A. (2000). The effect of 2G on mouse circadian rhythms. *Journal of Gravitational Physiology*, 7, 79–85.

Myerson, M.S. (2015). *Gastrocnemius, an Issue of Foot and Ankle Clinics of North America*. Elsevier, Amsterdam.

Najrana, T. and Sanchez-Esteban, J. (2016). Mechanotransduction as an adaptation to gravity [Online]. *Frontiers in Pediatrics*, 4. Available: https://doi.org/10.3389/fped.2016.00140.

Neumeister, H. and Budelmann, B.U. (1997). Structure and function of the nautilus statocyst [Online]. *Philosophical Transactions of the Royal Society of London Series B: Biological Sciences*, 352, 1565–1588. Available: https://doi.org/10.1098/rstb.1997.0142.

Niklas, K.J., Spatz, H.-C., and Vincent, J. (2006). Plant biomechanics: An overview and prospectus [Online]. *American Journal of Botany*, 93, 1369–1378. Available: https://doi.org/10.3732/ajb.93.10.1369.

Norman, T.L. and Wang, Z. (1997). Microdamage of human cortical bone: Incidence and morphology in long bones. *Bone*, 20, 375–379.

Oldeman, R.A.A. (1974). *L'architecture de la Forêt Guyanaise.* ORSTOM, Paris.

Oshima, M., Oshima, H., and Taketo, M.M. (2005). Hypergravity induces expression of cyclooxygenase-2 in the heart vessels [Online]. *Biochemical and Biophysical Research Communications*, 330, 928–933. Available: https://doi.org/10.1016/j.bbrc.2005.03.060.

Papaseit, C., Pochon, N., and Tabony, J. (2000). Microtubule self-organization is gravity-dependent [Online]. *Proceedings of the National Academy of Sciences*, 97, 8364–8368. Available: https://doi.org/10.1073/pnas.140029597.

Pedrozo, H.A., Schwartz, Z., Luther, M., Dean, D.D., Boyan, B.D., and Wiederhold, M.L. (1996). A mechanism of adaptation to hypergravity in the statocyst of aplysia californica. *Hearing Research*, 102, 51–62.

Perbal, G. and Driss-Ecole, D. (2003). Mechanotransduction in gravisensing cells [Online]. *Trends in Plant Science*, 8, 498–504. Available: https://doi.org/10.1016/j.tplants.2003.09.005.

Perbal, G., Lefranc, A., Jeune, B., and Driss-Ecole, D. (2004). Mechanotransduction in root gravity sensing cells [Online]. *Plant Physiology*, 120, 303–311. Available: https://doi.org/10.1111/j.0031-9317.2004.0233.x.

Pickering, R. and Kramers, J.D. (2010). Re-appraisal of the stratigraphy and determination of new U-Pb dates for the sterkfontein hominin site [Online]. *Journal of Human Evolution*, 59, 70–86. Available: https://doi.org/10.1016/j.jhevol.2010.03.014.

Piclet, B. (2009). Traitement chirurgical percutané des déformations des orteils latéraux. In *Chirurgie mini-invasive et percutanée du pied*, Cazeau, C. (ed.). Sauramps Médical, Montpellier, 269–291.

Piveteau, J., Lehman, J.-P., Dechaseaux, C., and Genet-Varcin, E. (1978). *Précis de Paléontologie des Vertébrés*. Masson, Paris.

Plieth, C. and Trewavas, A.J. (2002). Reorientation of seedlings in the earth's gravitational field induces cytosolic calcium transients [Online]. *Plant Physiology*, 129, 786–796. Available: https://doi.org/10.1104/pp.011007.

de Prado, M., Ripoll, P.L., and Golanó, P. (2005). *Cirugía Percutánea del Pie: Técnicas Quirúrgicas, Indicaciones, Bases Anatómicas*. Masson, Paris.

Preuschoft, H. (1971). Functional anatomy of the lower extremity. In *Immunology, Infections, Hormones, Anatomy, and Behavior, the Chimpanzee*, Carpenter, R.L. and Bourne, G.H. (eds). Karger, Basel.

Prieskorn, D., Graves, S., Yen, M., Ray, J., and Schultz, R. (1995). Integrity of the first metatarsophalangeal joint: a biomechanical analysis [Online]. *Foot & Ankle Society*, 16, 357–362. Available: https://doi.org/10.1177/107110079501600608.

Prisk, G.K. (2014). Microgravity and the respiratory system [Online]. *European Respiratory Journal*, 43, 1459–1471. Available: https://doi.org/10.1183/0903 1936.00001414.

Remis, M.J. (2013). *The Gorilla Paradox: The Effects of Body Size and Habitat on the Positional Behavior of Lowland and Mountain Gorillas.* Springer, Berlin.

Riesenfeld, A. (1974). Metatarsal robusticity in primates and a few other plantigrade mammals [Online]. *Primates*, 15, 1–25. Available: https://doi.org/10.1007/ BF01749589.

Robinson, E.L. and Fuller, C.A. (2000). Gravity and thermoregulation: Metabolic changes and circadian rhythms [Online]. *Pflügers Archiv*, 441, 32–38. Available: https://doi.org/10.1007/s004240000329.

Robling, A.G., Castillo, A.B., and Turner, C.H. (2006). Biomechanical and molecular regulation of bone remodeling [Online]. *Annual Review of Biomedical Engineering*, 8, 455–498. Available: https://doi.org/10.1146/annurev. bioeng.8.061505.095721.

Rodionova, N.V. and Oganov, V.S. (2003). Changes of cell-vascular complex in zones of adaptive remodeling of the bone tissue under microgravity conditions [Online]. *Advances in Space Research*, 32, 1477–1481. Available: https://doi.org/10.1016/ S0273-1177(03)90384-X.

Ronca, A.E., Fritzsch, B., Alberts, J.R., and Bruce, L.L. (2000). Effects of microgravity on vestibular development and function in rats: Genetics and environment [Online]. *Korean Journal of Biological Sciences*, 4, 215–221. Available: https://doi.org/10.1080/12265071.2000.9647547.

Ronca, A.E., Fritzsch, B., Bruce, L.L., and Alberts, J.R. (2008). Orbital spaceflight during pregnancy shapes function of mammalian vestibular system [Online]. *Behavioral Neuroscience*, 122, 224–232. Available: https://doi.org/10.1037/0735-7044.122.1.224.

Rose, M.D. (1994). Quadrupedalism in some Miocene catarrhines [Online]. *Journal of Human Evolution*, 26, 387–411. Available: https://doi.org/10.1006/jhev.1994.1025.

Rubin, J., Murphy, T., Nanes, M.S., and Fan, X. (2000). Mechanical strain inhibits expression of osteoclast differentiation factor by murine stromal cells [Online]. *American Journal of Physiology-Cell Physiology*, 278, 1126-1132. Available: https://doi.org/10.1152/ajpcell.2000.278.6.C1126.

Sack, F.D. and Leopold, A.C. (1985). Cytoplasmic streaming affects gravity-induced amyloplast sedimentation in maize coleoptiles [Online]. *Planta*, 164, 56–62. Available: https://doi.org/10.1007/BF00391025.

Sakai, A. (2012). Anabolic action of PTH under the loading/unloading condition [Online]. *Clinical Calcium*, 22, 365–371. Available: https://doi.org/CliCa1203365371.

Sarmiento, E.E. (1994). Terrestrial traits in the hands and feet of gorillas. *American Museum Novitates*, 3091, 1–56.

Schaffler, M.B., Choi, K., and Milgrom, C. (1995). Aging and matrix microdamage accumulation in human compact bone. *Bone*, 17, 521–525.

Schöne, H. and Steinbrecht, R.A. (1968). Fine structure of statocyst receptor of astacus fluviatilis [Online]. *Nature*, 220, 184–186. Available: https://doi.org/10.1038/220184a0.

Schultz, A.H. (1930). The skeleton of the trunk and limbs of higher primates. *Human Biology*, 2(3), 303–438.

Scurfield, G. (1973). Reaction wood: Its structure and function: lignification may generate the force active in restoring the trunks of leaning trees to the vertical [Online]. *Science*, 179, 647–655. Available: https://doi.org/10.1126/science.179.4074.647.

Seeman, E. (2003). The structural and biomechanical basis of the gain and loss of bone strength in women and men [Online]. *Endocrinology and Metabolism Clinics of North America*, 32, 25–38. Available: https://doi.org/10. 1016/S0889-8529(02)00078-6.

Seinandre, É. (2011). *Les Origines de L'homme Avant et Après Lucy*. Larousse, Paris.

Sekulić, S.R., Lukač, D.D., and Naumović, N.M. (2005). The fetus cannot exercise like an astronaut: Gravity loading is necessary for the physiological development during second half of pregnancy [Online]. *Medical Hypotheses*, 64, 221–228. Available: https://doi.org/10.1016/j.mehy.2004.08.012.

Shi, D., Meng, R., Deng, W., Ding, W., Zheng, Q., Yuan, W., Liu, L., Zong, C., Shang, P., and Wang, J. (2010). Effects of microgravity modeled by large gradient high magnetic field on the osteogenic initiation of human mesenchymal stem cells [Online]. *Stem Cell Reviews and Reports*, 6, 567–578. Available: https://doi.org/10.1007/s12015-010-9182-x.

Singla, C.L. (1983). Fine structure of the sensory receptors of aglantha digitale (Hydromedusae: Trachylina). *Cell and Tissue Research*, 231, 415–425.

Smith, S.E., Landorf, K.B., Butterworth, P.A., and Menz, H.B. (2012). Scarf versus chevron osteotomy for the correction of 1-2 intermetatarsal angle in hallux valgus: a systematic review and meta-analysis [Online]. *Journal of Foot & Ankle Surgery*, 51, 437–444. Available: https://doi.org/10.1053/j.jfas.2012.02.016.

Solé, M., Lenoir, M., Durfort, M., López-Bejar, M., Lombarte, A., and André, M. (2013). Ultrastructural damage of loligo vulgaris and illex coindetii statocysts after low frequency sound exposure [Online]. *PLoS ONE*, 8. Available: https://doi.org/10.1371/journal.pone.0078825.

Steinböck, G. (1996). Austin's horizontal V-shaped sliding osteotomy of the metatarsal head (Chevron-osteotomy) in the treatment of hallux valgus. *Orthopade*, 25, 308–316.

Stienstra, J.J., Lee, J.A., and Nakadate, D.T. (2002). Large displacement distal chevron osteotomy for the correction of hallux valgus deformity. *Journal of Foot & Ankle Surgery*, 41, 213–220.

Suero, E.M., Meyers, K.N., and Bohne, W.H.O. (2012). Stability of the metatarsophalangeal joint of the lesser toes: A cadaveric study [Online]. *Journal of Orthopaedic Research*, 30, 1995–1998. Available: https://doi.org/10.1002/jor.22173.

Swift, J.M., Lima, F., Macias, B.R., Allen, M.R., Greene, E.S., Shirazi-Fard, Y., Kupke, J.S., Hogan, H.A., and Bloomfield, S.A. (2013). Partial weight bearing does not prevent musculoskeletal losses associated with disuse [Online]. *Medicine & Science in Sports & Exercise*, 45, 2052–2060. Available: https://doi.org/10.1249/MSS.0b013e318299c614.

Tabony, J., Rigotti, N., Glade, N., and Cortès, S. (2007). Effect of weightlessness on colloidal particle transport and segregation in self-organising microtubule preparations [Online]. *Biophysical Chemistry*, 127, 172–180. Available: https://doi.org/10.1016/j.bpc.2007.01.010.

Takahashi, K., Okumura, H., Guo, R., and Naruse, K. (2017). Effect of oxidative stress on cardiovascular system in response to gravity [Online]. *International Journal of Molecular Sciences*, 18, 1426. Available: https://doi.org/10.3390/ijms18071426.

Talbot, J.M. and Fisher, K.D. (1986). Influence of space flight on red blood cells. *Federation Proceedings*, 45, 2285–2290.

Tamm, S.L. (2014). Formation of the statolith in the ctenophore mnemiopsis leidyi [Online]. *The Biological Bulletin*, 227, 7–18. Available: https://doi.org/10.1086/BBLv227n1p7.

Thompson, D.W., Teyssié, D., Bonner, J.T., Gould, S.J., and Prochiantz, A. (2009). *Forme et croissance*. Le Seuil, Paris.

Trinkaus, E. (2005). Anatomical evidence for the antiquity of human footwear use [Online]. *Journal of Archaeological Science*, 32, 1515–1526. Available: https://doi.org/10.1016/j.jas.2005.04.006.

Trinkaus, E. and Shang, H. (2008). Anatomical evidence for the antiquity of human footwear: Tianyuan and sunghir [Online]. *Journal of Archaeological Science*, 35, 1928–1933. Available: https://doi.org/10.1016/j.jas.2007.12.002.

Turko, A.J., Kültz, D., Fudge, D., Croll, R.P., Smith, F.M., Stoyek, M.R., and Wright, P.A. (2017). Skeletal stiffening in an amphibious fish out of water is a response to increased body weight [Online]. *The Journal of Experimental Biology*, 220, 3621–3631. Available: https://doi.org/10.1242/jeb.161638.

Turner, C.H., Forwood, M.R., and Otter, M.W. (1994). Mechanotransduction in bone: Do bone cells act as sensors of fluid flow? [Online]. *The FASEB Journal*, 8, 875–878. Available: https://doi.org/10.1096/fasebj.8.11.8070637.

Turner, C.H., Owan, I., and Takano, Y. (1995). Mechanotransduction in bone: Role of strain rate [Online]. *American Journal of Physiology*, 269, 438–442. Available: https://doi.org/10.1152/ajpendo.1995.269.3.E438.

Tuttle, R.H. and Watts, D.P. (1985). The positional behavior and adaptive complexes of pan gorilla. In *Primate Morphophysiology, Locomotor Analyses and Human Bipedalism*, Kondo, S. (ed.). University of Tokyo Press, Tokyo, 261–288.

Vico, L., Chappard, D., Alexandre, C., Palle, S., Minaire, P., Riffat, G., Morukov, B., and Rakhmanov, S. (1987). Effects of a 120 days period of bed-rest on bone mass and bone cell activities in man: Attempts at countermeasure. *Bone Miner*, 2, 383–394.

Vinnikov, Y.A., Gazenko, O.G., Titova, L.K., Bronstein, A.A., Tsirulis, T.P., Pevzner, R.A., Govardovskii, V.I., Gribakin, F.G., Pal'mbakh, L.P., Aronova, M.Z., Mashinskii, A.L., Ivanov, V.P., Kharkeevich, T.A., and Pyatkina, G.A., (1974). The role of gravity in the phylogeny of structure and function in animal sensors of spatial orientation, and their predicted action in weightlessness. *Life Sciences in Space Research*, 12, 159–176.

Volkmann, D. and Baluška, F. (2006). Gravity: one of the driving forces for evolution [Online]. *Protoplasma*, 229, 143–148. Available: https://doi.org/10.1007/s00709-006-0200-4.

Volkmann, D., Baluska, F., Lichtscheidl, I., Driss-Ecole, D., and Perbal, G. (1999). Statoliths motions in gravity-perceiving plant cells: does actomyosin counteract gravity? [Online]. *FASEB Journal*, 13, 143–147. Available: https://doi.org/10.1096/fasebj.13.9001.s143.

Von Meyer, G.H. (2011). The classic: The architecture of the trabecular bone (tenth contribution on the mechanics of the human skeletal framework) [Online]. *Clinical Orthopaedics and Related Research*, 469, 3079–3084. Available: https://doi.org/10.1007/s11999-011-2042-4.

Vovchuk, S.V., Barabash, R.D., and Levitsky, A.P. (1975). Isolation and properties of 2 kallikreins from human saliva. *Voprosy meditsinskoĭ khimii*, 21, 622–625.

Walton, K.D., Lieberman, D., Llinás, A., Begin, M., and Llinás, R.R. (1992). Identification of a critical period for motor development in neonatal rats [Online]. *Neuroscience*, 51, 763–767. Available: https://doi.org/10.1016/0306-4522(92)90517-6.

Warden, S.J., Galley, M.R., Richard, J.S., George, L.A., Dirks, R.C., Guildenbecher, E.A., Judd, A.M., Robling, A.G., and Fuchs, R.K. (2013). Reduced gravitational loading does not account for the skeletal effect of botulinum toxin-induced muscle inhibition suggesting a direct effect of muscle on bone [Online]. *Bone*, 54, 98–105. Available: https://doi.org/10.1016/j.bone.2013.01.043.

Weidenreich, F. (1923). Evolution of the human foot [Online]. *American Journal of Physical Anthropology*, 6, 1–10. Available: https://doi.org/10.1002/ajpa.1330060102.

Weinbaum, S., Zhang, X., Han, Y., Vink, H., and Cowin, S.C. (2003). Mechanotransduction and flow across the endothelial glycocalyx [Online]. *Proceedings of the National Academy of Sciences*, 100, 7988–7995. Available: https://doi.org/10.1073/pnas.1332808100.

Whalen, R. (1993). Musculoskeletal adaptation to mechanical forces on Earth and in space. *Physiologist*, 36, 127–130.

Whitson, P.A., Putcha, L., Chen, Y.-M., and Baker, E. (1995). Melatonin and cortisol assessment of circadian shifts in astronauts before flight [Online]. *Journal of Pineal Research*, 18, 141–147. Available: https://doi.org/10.1111/j.1600-079X.1995.tb00152.x.

Wiederhold, M.L., Harrison, J.L., Parker, K., and Nomura, H. (2000). Otoliths developed in microgravity. *Journal of Gravitational Physiology*, 7, 39–42.

Wilson, J.N. (1963). Oblique displacement osteotomy for hallux valgus. *Journal of Bone and Joint Surgery*, 45, 552–556.

Wilson, B.F. and Archer, R.R. (1979). Tree design: Some biological solutions to mechanical problems [Online]. *BioScience*, 29, 293–298. Available: https://doi.org/10.2307/1307825.

Winter, D.A. and Yack, H.J. (1987). EMG profiles during normal human walking: Stride-to-stride and inter-subject variability. *Electroencephalography and Clinical Neurophysiology*, 67, 402–411.

Wubbels, R., Van Marle, J., Sondag, H.N.P., and de Jong, H.A. (2002). Effects of hypergravity on the morphological properties of the vestibular sensory epithelium, II. Life-long exposure of rats including embryogenesis [Online]. *Brain Research Bulletin*, 58, 575–580. Available: https://doi.org/10.1016/S0361-9230(02)00828-6.

Wunderlich, R.E. (1999). *Pedal form and Plantar Pressure Distribution in Anthropoid Primates*. Stony Brook University, New York.

Wynes, J., Lamm, B.M., Andrade, B.J., and Malay, D.S. (2016). Preoperative planning and intraoperative technique for accurate translation of a distal first metatarsal osteotomy [Online]. *Journal of Foot & Ankle Surgery*, 55, 49–54. Available: https://doi.org/10.1053/j.jfas.2015.05.009.

Xie, L.Q. and Liu, C.L. (1999). Developments of the theory of skeletal adaptation to mechanical loading. *Space Medicine Medical & Engineering*, 12, 226–230.

You, L., Cowin, S.C., Schaffler, M.B., and Weinbaum, S. (2001). A model for strain amplification in the actin cytoskeleton of osteocytes due to fluid drag on pericellular matrix. *Journal of Biomechanics*, 34, 1375–1386.

Yuge, L., Sasaki, A., Kawahara, Y., Wu, S., Matsumoto, M., Manabe, T., Kajiume, T., Takeda, M., Magaki, T., Takahashi, T., Kurisu, K., and Matsumoto, M. (2011). *Stem Cells and Development*, 20, 893–900. Available: https://doi.org/10.1089/scd.2010.0294.

Index

A, B, C

D, E, F

G, H, I

J, K, L

M, N, O

P, Q, R

S, T, U

V, W, Y, Z

Other titles from

iSTE

in

Biology and Biomedical Engineering

2020

DUJON Bernard, PELLETIER Georges
Trajectories of Genetics

2019

BRAND Gérard
Discovering Odors

BUIS Roger
Biology and Mathematics: History and Challenges

2016

CLERC Maureen, BOUGRAIN Laurent, LOTTE Fabien
Brain-Computer Interfaces 1: Foundations and Methods
Brain-Computer Interfaces 2 : Technology and Applications

FURGER Christophe
Live Cell Assays: From Research to Health and Regulatory Applications

2015

CLARYSSE Patrick, FRIBOULET Denis
Multi-modality Cardiac Imaging: Processing and Analysis

2014

CHEZE Laurence
Kinematic Analysis of Human Movement

CLARYSSE Patrick, FRIBOULET Denis
Multi-modality Cardiac Imaging: Processing and Analysis

DAO Tien Tuan, HO BA THO Marie-Christine
Biomechanics of the Musculoskeletal: Modeling of Data Uncertainty and Knowledge

FANET Hervé
Medical Imaging Based on Magnetic Fields and Ultrasounds

FARINAS DEL CERRO Luis, INOUE Katsumi
Logical Modeling of Biological Systems

MIGONNEY Véronique
Biomaterials

TEBBANI Sihem, LOPES Filipa, FILALI Rayen, DUMUR Didier, PAREAU Dominique
CO_2 Biofixation by Microalgae: Modeling, Estimation and Control